Let's Grow Tomatoes

Let's Grow Tomatoes

The Grow-Box Method to Bountiful Harvests

Dr. Jacob R. Mittleider

COPYRIGHT © 1981 BY
HORIZON PUBLISHERS

All rights reserved. Reproduction in whole or any
parts thereof in any form or by any media without
written permission is prohibited.

HORIZON PUBLISHERS CATALOG AND ORDER NUMBER
4027

INTERNATIONAL STANDARD BOOK NUMBER
0-88290-176-1

LIBRARY OF CONGRESS CATALOG CARD NUMBER
80-84563

**Horizon
Publishers &
Distributors**

P.O. Box 490
50 South 500 West
Bountiful, Utah 84010

Dedication

This book is dedicated to Dr. and Mrs. Dee L. Stoops. Though skilled in the medical sciences, they couldn't resist growing tomatoes. Their enthusiasm and positive attitude in facing successes and failures were contagious—it was the stimulus which kept me constantly experimenting with the world's most popular vegetable, "tomatoes." This publication is the result of that work, and it came into being because of the confidence shown to me by this wonderful couple.

J. R. Mittleider

Dr. Jacob R. Mittleider
International Agriculture Consultant

Contents

Color Photo Section 9
Introduction 17
Importance of the Tomato 18
What Are Grow-Boxes? 20
    Selecting and Preparing the Area 21
    Tools Needed to Build Wooden Frames 23
    Materials Needed to Build One 5' × 30' × 8" Grow-Box 23
    Construction Materials 24
    How to Construct a Grow-Box 24
    Several Factors to Remember When Building Frames 29
Grow-Box Soils 30
How to Get Plants! 33
    Advantages in Growing Plants from Seed 34
Facts to Know About Seed 36
    Seed Viability 36
    Treating the Seed Against Disease 36
    Sterilizing Soil for Seed 38
    Problems Found in Using Regular Field Soil 38
    The Weed Problem 38
    Starting Plants From Seed 39
    The Preplant Fertilizer Formula 39
    How Many Seeds Per Seedflat? 41
Transplanting Seedlings 44
    Filling Square Pots 45
    Planting Instructions 45
Constant Feed Method 48
    The Mittleider Weekly Feeding Formula 48
    How to Apply the Solution 49
How to Keep Plants From Growing Spindly 51
    Preventing Plants From Growing Spindly 52
Producing Large-Size Plants 54
    Shift Plants into Larger-Size Containers 54
    Additional Information on Gallon-Size Containers 57
Pruning, Staking, Tying 60
    Suckers and Pruning 60
    Staking and Tying 61
    Flower Set 62
    Leaves 63
Managing Single-Stem Vines 64
    The "Stake" Method 65
    Installing Stakes 66
    The "A" Frame Method 67
    How to Fasten the Strings to the "A" Frames 69
Spacing Tomato Plants in Grow-Boxes 71
    Calculating the Number of Plants One Grow-Box Holds 72
    What Size Plants Should Be Transplanted? 72
    Tips on Mixing Crops in Grow-Boxes 72
Transplanting Procedures Illustrated 75
    How to Plant Extra-Long Tomato Plants 76

8 LET'S GROW TOMATOES

Pruning Fruiting Tomatoes 79
    Caution When Diagnosing Symptoms 81
    More Information on Leaves and Suckers 83
Daily and Weekly Care 87
Watering Tomatoes 89
    The Goal in Watering 90
    Water Requirements for Tomatoes 92
Fertilizers 93
    Feeding Established Plants 94
    The Weekly Feeding Procedure 96
    Actual Fertilizing Procedures 98
    A Review of Feeding Instructions 100
    When to Decrease the Fertilizers 100
Flower and Fruit Set 102
    The Light Factor 103
    The Heat Factor 103
    The Plant Population 103
    Cannery Tomatoes 104
    Market and Fruit Stands 104
    Greenhouse Tomato Crops 104
    The Family Garden 105
Nematodes 107
    Root Nematodes (Eel-like Worms) 107
    How to Inspect for Nematode Infestation 108
    How Plants React to Nematodes 108
Fertilizer and Soil Problems 110
Insects and Soil Maggots 113
    Insects 113
    Other Factors to Look At When the Fruit Drops 114
    Checking For Soil Maggots 115
    Diagnosing for Soil Maggots 116
    Treatment to Control Soil Maggots 117
Plant Diseases 119
    Greenhouse Crops an Exception 119
    "Curley-Top" Disease 120
    "Early and Late Blight" Disease 121
    "Virus" Disease 122
Weather Problems and Vibrating Plant Vines 125
    Shaking and Vibrating Tomato Vines 125
    Summary 126
Cracked Fruit 127
Blossom-End Rot On Tomatoes 129
    Blossom-End Rot Caused from Stress 129
    Blossom and Stem-End Rot Caused from Infection 129
Thinning Tomatoes 132
Harvesting Tomatoes 133
    Lengthening the Ripe-Tomato Harvest Season 134
Appendix 1 137
Appendix 2 139
Appendix 3 140
Index 141

Tomatoes are a vital ingredient in any gardening project.

Grow-boxes can be equipped with watering systems which increase the yield and reduce man-hours of work.

Tomatoes grown in pots in homemade greenhouses add several weeks to the growing season.

Transplanting tomato plants into a mini grow-box.

Concrete forms are durable grow-box frames.

Preparing grow-boxes for a new planting.

Stores can't match the flavor of the home gardener's succulent vine-ripened tomatoes.

Components of a watering system for mini grow-boxes.

The Mittleider grow-box system yields firm, vine-ripe tomatoes from plants seven feet tall.

16 LET'S GROW TOMATOES

A careful spraying program protects against animal invaders.

Careful pruning is the key to firm tomatoes of uniform size and quality.

1

Introduction

Have you ever wondered what it would be like not to have tomatoes? No doubt we would survive, but something wonderful would be missing!

Tomatoes are hardy plants that will grow under a wide range of conditions. But even though they are easy to grow, the crops are sometimes disappointing and unprofitable because the yields are small and the shape and quality of the fruit is poor.

Uniform size and high quality crops require skill and perseverance on the part of the gardener or farmer.

Production methods vary, depending on what the crop is used for—marketing, canning, or for table use.

The steady increase in the fixed costs of production, and the price the retail public is willing to pay for market-ripe tomatoes, places constant pressure on the grower. In order for him to survive, he must increase production yields, improve the quality of the fruit and reduce his fixed operating costs.

During the past few years, great changes have been introduced in the growing and production of tomatoes. Among the changes, one that has repeatedly demonstrated its worth is the Mittleider Grow-Box Method.

The history of growing crops in grow-boxes dates back about 40 years. But only recently has the national and international interest in food production attracted sufficient attention to investigate the enormous potential of the grow-box method. Many years of research have been spent in grow-box crops.

This publication deals with factors associated with tomato production specifically in grow-boxes. However, the methods outlined can be used with success in various other methods of growing.

2

The Importance Of The Tomato

To people around the world, the word "tomato" is like a bit of sweet music. In some languages it is called the "love-apple," and the name is quite fitting!

The dictionary defines the tomato as "a widely cultivated solanaceous (Solanum, meaning a large genus) family of plants, bearing a slightly acid, pulpy fruit, commonly red, sometimes yellow; the fruit itself is used as a vegetable."

The family name for this type of plant is "Lycopersicon." In this family there are several varieties of important domesticated crops and some uncultivated varieties—some good, some bad. For example:

- *Potato,* an edible tuber, called "ground berry" in the German language.
- *Tomato,* a widely-cultivated plant. There are many varieties, such as the cherry tomato and the Big Boy.
- *Peppers,* the annual sweet and hot varieties.
- *Eggplant,* several varieties (annual and bi-annual).
- *Tobacco,* A poisonous herb with narcotic leaves.
- *Nightshade weeds,* several varieties which have white flowers and bear poisonous berries. And,
- *Jimson weed,* a coarse ill-smelling weed with white flowers, poisonous spiny fruit and narcotic leaves.

From this partial list it can be seen that the tomato has an impressive heritage. And fortunately, the tomato and its relatives are not restricted by climatic zones—they will grow in the tropics, the sub-tropics, the temperate, and cold-temperate climates. However, they will freeze at 32°F.

The tomato is treated as an annual (planted every year), but under some conditions the same plants will live two years and sometimes longer.

THE IMPORTANCE OF THE TOMATO

In America, until about 1935, field-crop tomatoes were comparatively easy to grow. They produced fruit in almost any garden soil with minimum care. The "Old Timers" still reminisce about picking a bushel (60 pounds) of large, ripe, juicy fruit from a single tomato plant.

Since 1935, the problems associated with growing tomatoes have been increasing in many areas of production. Through continued research, new disease and virus-resistant varieties have been developed. They are being introduced rather frequently, and the tomato continues to be very popular.

No garden seems to be complete unless it has a few tomato vines. And since this is so, this publication has been prepared to help you produce more tomatoes, in less space, and to harvest ripe fruit over a longer season.

3

What Are Grow-Boxes?

Grow-boxes are small garden plots enclosed in frames which are leveled in place and filled with special custom-made soil. They are used primarily to produce special crops for commercial purposes and fresh vegetables for family use.

The frames can be made of many kinds of non-toxic materials. The grow-boxes can vary in size, but the most common size is 5 feet wide by 30 feet long by 8 inches deep.

A grow-box frame.

WHAT ARE GROW-BOXES? 21

Grow-boxes rest on top of the soil surface, and they have no bottoms or lids. They can be constructed almost anywhere—on steep hillsides, over rocks, over alkali or clay soils, eroded soils, over swampy land, and even on a carport. But remember the frames must be level!

Hillside frames.

Selecting and Preparing the Area

1. The size of the area depends on the land available and the number of grow-boxes planned.
2. Almost all vegetable crops require sunshine—choose a sunny location.
3. Stay away from shade caused by trees and hedges and buildings.
4. The ideal is for full sunshine on the grow-box throughout the entire day.
5. Winds, whether cold or hot, injure the leaves and crops. Avoid windy areas and choose protected garden spots.
6. Stay out of low areas, especially where water collects or stands.

7. Build on high ground. This is primarily to assure unrestricted drainage.

8. Plants, too, get thirsty! Rains are usually not dependable as a source of water throughout all the year. Build grow-boxes close to a reliable source of water.

9. Walkways between and around grow-boxes are necessary. Make them safe for walking, either by removing all obstacles or by covering them with sand or soil.

10. For ease and convenience, include an access road into the area.

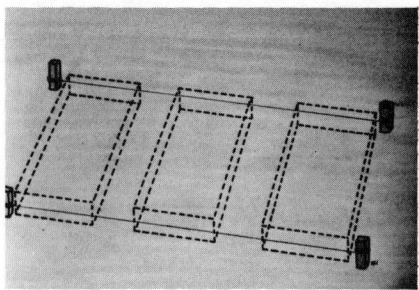

Make a sketch of the area.

Avoid shade.

Sunshine is needed all day.

Build on high ground.

Now, mark off and level the whole area, if possible. Or, mark off plots for each grow-box if the area is a hillside. Remove weeds and brush, rocks, vines, dead logs, etc.

WHAT ARE GROW-BOXES?

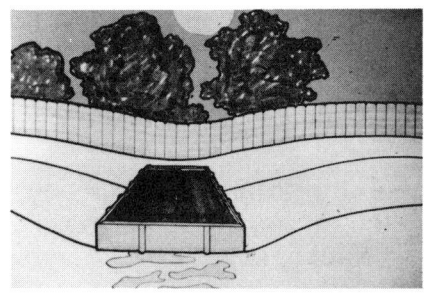

Avoid low areas with poor drainage.

Have water handy.

Make aisles safe for walking.

A pickup lightens the work.

Tools Needed to Build Wooden Frames

- 1 claw hammer
- 1 ball chalkline or ball nylon string
- 1 handsaw or skil saw
- 1 4-pound hammer to drive wooden stakes
- 1 level (2 feet long or longer)
- 1 shovel for digging and moving dirt

Materials Needed to Build One 5' × 30' × 8" Grow-Box

- 70 feet 1" × 8" × 10' redwood or cedar lumber
- 24 1" × 2" × 18" redwood or cedar stakes—pointed at one end
- 1 pound 3" nails (box type)

Construction Materials

Grow-boxes are usually made of lumber or cement. But other materials such as bricks, cement blocks, rocks, small-diameter straight poles, metal, etc., can be used.

Please note: Some soft-wood lumber rots quickly unless it is treated with a preservative. Paint and copper-based compounds are usually safe to treat lumber against decay. But *do not* use kreosote. *Kreosote is toxic to plants.*

A cement frame grow-box.

Frames can be made with bricks, blocks, etc.

How to Construct a Grow-Box

1. Drive two stakes 35 feet apart and stretch a string between the stakes.
2. Drive two more stakes 10 feet apart at right angles to the first two stakes and stretch a line between them.
3. Level the area, at least partially, under the strings.

Stretching the first chalkline.

Keep the chalkline tight.

WHAT ARE GROW-BOXES?

Level the area partially.

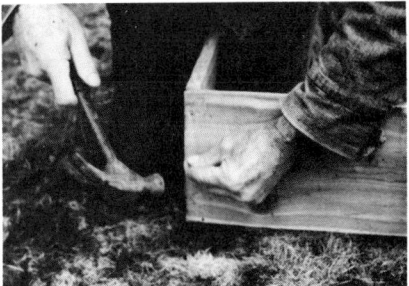

Starting the grow-box frame.

4. Take one long board and one end board and nail the ends together.

5. Move the boards into position—the point where the two strings cross.

6. Drive the first stake in line with the longest string, and back about 8 inches from the end of the grow-box.

7. Place the level on the top edge of the long board.

8. Raise or lower the board to make it level.

9. When level, stake and nail the *stake* to the board.

10. Bend the end of the nail downward against the board inside the box frame.

11. Next, move down the board 6 to 8 feet and drive a stake in line with the string. Level the board and nail the stake to the board.

12. Move to the end of the board. This time half of the stake should extend beyond the end of the board. This is necessary to splice the next length of board. Level the stake and nail it.

Setting the frame in position.

Drive the first stake 8 inches from the end.

26 LET'S GROW TOMATOES

Checking for level.

Leveling the frame.

Nail the stake to the board.

Bend the nail downward.

Level the whole box frame.

Nail two boards to same stake.

13. There are two kinds of splices to choose from Both are satisfactory!

14. Continue the process of leveling, staking and nailing to the end of the grow-box.

15. Next, drive a stake along the end string. Level, stake and nail the end board.

WHAT ARE GROW-BOXES?

16. To level the end of the grow-box and the opposite side, place the level across the corner of the frame. When level, stake and nail.

17. A simple aid used for leveling and for lining up the opposite side of the grow-box frame is the use of the spreader board and the level.

17. Leveling across the corners gives a more accurate level than leveling from the top edge of the board.

18. To strengthen the grow-box frame, drive additional stakes and nail them to the frame.

19. When the grow-box is properly leveled, staked and nailed, it looks like this!

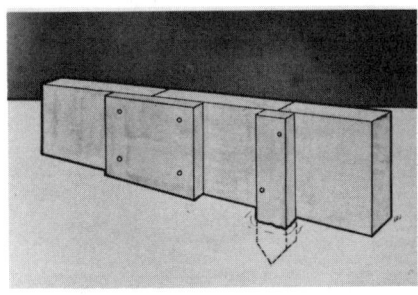

Two kinds of splices.

Staking the frame.

Staking the end board.

Leveling the end board.

Using a spreader board.

Leveling the corner.

Adding extra stakes.

View of a completed frame.

Several Factors to Remember When Building Frames

- Always drive the stakes on the outside of the frames.
- Nail the stake to the boards—not visa versa.
- The frames, both length and width, must be level.
- The tops of the stakes and frames should be even.
- When treating lumber, do not use kreosote!
- The frames must rest on top of the ground. Do not sink them into the ground.

For more complete details and explanations and drawings on grow-box construction, grow-box soils, grow-box location, filling grow-boxes with special soil, etc., refer to the book *More Food From Your Garden,* by the same author.

4

Grow-Box Soils

Experience with various crops grown in grow-box soil demonstrated that cold-weather plants grew stronger and sturdier in cold weather and did well even into hot weather. The experiments showed also that heat-loving plants did better in the heat of summer, when the plants grew in custom-made soil, as compared to growing in regular soils of the field and garden.

What made the difference? The grow-box soil is fluffy, lightweight, porous, loose, yet stable enough to resist wind and water erosion. It allows water to penetrate (percolate) freely. The soil can be tilled easily and quickly with simple garden tools, or just by hand. It eliminates the need for powered equipment. Because the boxes are level, it is easy to fertilize evenly and accurately, and abundant oxygen exchange to the roots of the plants is allowed.

The structure of the soil provides a relatively even temperature in which cool-weather plants grow well, even in the heat of summer.

The few persistent weed seeds, which blow in and germinate occasionally, can be destroyed easily merely by rubbing the hand over the surface of the grow-box soil surface between the plants.

When vegetable crops are grown in grow-boxes, there is no reason to abandon the project to the weeds!

The soil combinations recommended for sprouting seeds or for filling the grow-boxes are a mixture of: sand and peat moss; sawdust and sand; sand, peat moss and perlite; coffee hulls and fine sand, or other similar materials. Please note, however, that regular garden and field soils are *not* recommended to fill grow-boxes!

Seeds which are sprouting require oxygen, moisture, and warmth. The soil combinations recommended above meet these specifications.

Note also that the list is only part of the materials which are generally available. Before rejecting or adopting any product to use as a soil media, the following questions should be considered:

1. Is it toxic to plants?
2. Does it decompose (decay) within four months? The time is important!
3. Does the material repel moisture?
4. Does it heat and ferment?
5. Does it settle into layers, pack and harden as the plants grow?

Custom-made soil.

Suggested soil combinations.

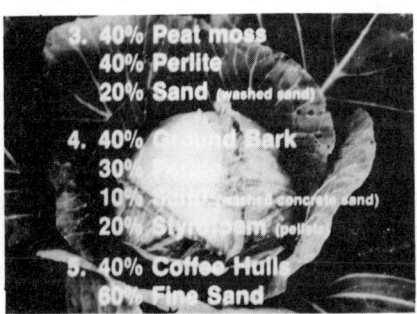

Suggested soil combinations.

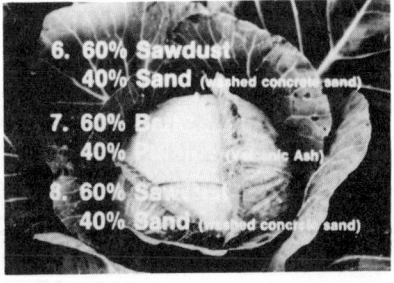

Suggested soil combinations.

If the answer to any of the questions is "yes," the products should not be used. If the answers are "no," the materials are usually safe to use.

The materials listed above are much improved if two or more are combined. The following combinations are recommended:

- 40% sawdust, 40% perlite, 20% medium or fine sand.
- 45% coffee hulls, 55% pumice.
- 40% peat moss, 30% perlite, 30% medium or fine sand.
- 45% pine bark, 20% perlite, 35% sand.
- 65% sawdust, 35% blowsand.

When combining the materials, measure them by volume, not by weight. Also, the different percentages of the materials can be increased or decreased. And with many materials to choose from, it is well to select from those which are locally available and cheap in price.

The materials chosen can be mixed together by hand, with a cement mixer, or with a tractor that has a front-end loader attachment. They should be mixed before filling the grow-boxes.

For small projects, however, the materials can be spread out in layers in the boxes and then mixed together.

A group mixing the special soil.

Hand mixing in the grow-box is adequate.

5

How To Get Plants!

There are two ways to produce tomato plants: One is by planting seed; the other is through rooting cuttings taken from growing plants.

The science of sprouting seeds and growing strong seedlings is explained in the chapters which follow, and everyone can choose whether to grow plants from seed or buy the plants from an established grower when they are needed.

In either case, it is recommended that the plants be 4 to 8 weeks old and hardened off—exposed to full sun and air at least 72 hours—before they are transplanted where they will bear fruit.

Seed catalogs list many varieties of tomatoes. Nearly all produce red fruit, but several varieties produce yellow fruit. The yellow fruit is equal in flavor with the red fruit and the plants are just as easy to grow.

The varieties differ in the number of days required to produce ripe fruit from seed. Some take 52 days, some 67 days, some 90 to 110 days.

Sometimes just looking over the large list of tomato varieties listed in a seed catalog and the variable number of days required to produce ripe fruit leads to confusion over how to make the best choices in selecting seed.

This problem can be simplified, however, if the varieties most commonly grown in your area are utilized for the main crop varieties, and only one or two new varieties are planted each year on an experimental basis. This procedure can prove rewarding in a relatively short while in determining which varieties do best in specific areas. To a large degree, the variety should be chosen on the basis of what the crop will be used for and on the type of propagating equipment available.

The answers to the following questions may be helpful in determining which are the best varieties to grow:

1. Are you growing plants from seed to produce tomatoes for your own needs and for marketing?
2. Do you have a cold-frame or a box-type greenhouse to sprout seeds and grow seedlings?
3. Or, do you have a genuine greenhouse, equipped to grow seedlings during cold weather?

Obviously, if adequate facilities for growing plants from seed are *not* available, the plants should be purchased from one who has the facilities.

When the facts are all considered, it is usually more economical for the average family to buy the tomato plants for the garden at the time they are needed, than it is to grow them. However, to depend on the market for plants carries an element of risk and other limiting factors. For example:

1. Most gardening people get the urge to plant gardens at the same time. Therefore, the larger, well-hardened plants are quickly bought and disappear from the market. This leaves the softer and very small plants.
2. The younger the seedlings, the longer it takes to pick vine-ripe tomatoes. Using very young seedlings consequently shortens the picking season before the frost kills the crop. This means a reduction in yield, in sales, and in the potential income from the crop.

For these and other reasons, the trend to grow plants from seed, for private use, is expected to increase.

The information which follows is *specifically* on tomato production and deals with some of the basic factors associated with seed germination and plant care throughout the entire growing season.

Advantages in Growing Plants From Seed

There are two major advantages to growing plants from seed. The first is that because tomatoes are everbearing, the labor and expense involved in growing the crop is nearly the same whether harvesting lasts three weeks or three months. And it is possible when growing plants from seed to lengthen the picking season as much as twelve weeks. To illustrate, if the normal planting season in the field or garden is May 15, the seed can be planted in the nursery February 15.

The procedures for growing sturdy seedlings in the nursery for a lengthy period of time are explained in the chapters which follow.

The second advantage is that when growing plants from seed, a larger variety can be grown. And by planting several varieties, vine-ripe tomatoes can be picked over a longer period of time. For example, if varieties like "pixey" which produce vine-ripe tomatoes in 52 days, "Earlians" which ripen fruit in 65 days, and "Better Boy" which ripen fruit in 80 to 110 days are grown, picking vine-ripe tomatoes can begin 52 days after planting seed and can continue right on through the summer and fall seasons—in fact, until the frost kills the plants.

6

Facts To Know About Seed

Seed Viability

Under normal conditions, tomato seed remains viable (fertile) for two to ten years in arid and dry regions. In contrast, in humid and tropical areas, unless the seed is stored with the fresh vegetables in the family refrigerator, the seed will die in 30 days!

But even though tomato seed does have a long life potential, there is really no economy in buying more seed than will be used in one or two seasons.

Treating the Seed Against Disease

Many vegetable seeds sold today are treated against disease and are packaged in vacuum-sealed packets or tins. It is safe to plant them without further treatment for disease control. Seeds which have not been treated previously carry a high risk as potential disease carriers. To be sure such seeds are not diseased, they should be heat-treated with hot water before planting.

The recommended treatment for tomato seeds is to immerse the seeds in 130°F. water for 30 minutes. Cool the seeds quickly after treatment and dry them thoroughly in the air. Heat-treated seeds can be planted immediately after treatment, or stored for planting weeks or months later. Be aware that hot water treatment of tomato seeds reduces germination by about 10%. The heat kills the weaker seeds.

FACTS TO KNOW ABOUT SEED

Do not put heat-treated seeds back into the original containers! Place them in new envelopes or plastic containers. Putting treated seed back into the original containers will re-contaminate the seed! For more detailed information on hot-water treatment of plants and seeds, refer to the book *Food For Everyone,* by Mittleider and Nelson, chapters 52 and 58.

A valuable reference book.

A tray of tomato seedlings.   *Pots of seedlings.*

Sterilizing Soil for Seed

Tomato seed can be sprouted in three days, or it may take two to three weeks. Temperature and moisture make the difference. Tomatoes are heat-loving plants and the seed germinates rapidly in soil that is uniformly warm—between 70 and 80°F.

If only a few plants are needed, seed can be sown broadcast in a 4-inch plastic pot or in a narrow tray. The container should be clean, free from disease, and filled with soft sterilized soil. If regular garden or field soil is used to sprout seed, it should be sterilized previously, even if sterilizing is done in the kitchen oven, to be sure it is disease free.

Sterilizing soil in the family oven is a quick operation. Spread the desired amount of soil ½-inch deep on a flat metal cookie sheet. Place it in the oven and heat to 250° for 30 minutes. Remove the soil from the oven and use when cool. Or, if stored for use later, put the soil in a clean container. To avoid re-contamination, store it in containers which have tight lids.

Problems Found in Using Regular Field Soil

Nearly all field and garden soils are depleted in humus, which is the remains from decomposed organic residues. Consequently, the soils develop wide cracks as they dry and set hard like cement. Water penetration (percolation) to the roots is slow and restricted. The oxygen supply in the soil is inadequate and stale. Machinery, which is expensive to purchase and costly to operate, is necessary to prepare the hard soil for planting and growing crops.

The Weed Problem

Weeds are plants out of place. They are highly-specialized plants which thrive even in poor soils and luxuriate in soils prepared for gardens.

Often when gardens are grown in regular soil they are planted and cared for enthusiastically for the first several weeks of the season and later neglected. The weeds quickly overrun the vegetable crops and the results for the hard work are disappointing and discouraging.

FACTS TO KNOW ABOUT SEED

Weed control problems, plus the struggle and the expenses usually encountered in gardening on regular soils, are factors which prompted the research and experimentation to overcome the problems. The grow-boxes and the special soil recommended are the results of the research projects.

Starting Plants From Seed

Here's how to start tomato plants from seed.

Plant the seed 8 to 12 weeks before the danger of frost is past and it is warm enough to plant in the garden. Suggested dates are February 9 to 16.

Step one: Fill a seedflat, 4-inch pot, or narrow tray with special soil, such as 75% peat moss, 25% perlite or sand.

Regular seedflats are 18 inches square and 3 inches deep, outside dimensions. They have bottoms with cracks for drainage, but no lids.

Step two: Sprinkle evenly 1½ *ounces* of the "Preplant Fertilizer" over each seedflat, and mix it thoroughly with the soil.

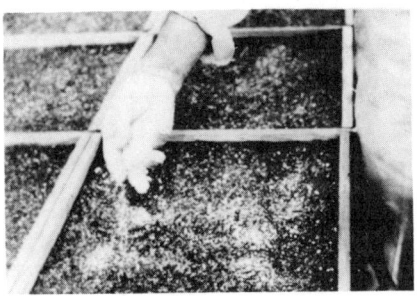

An empty standard-size flat.   *Spreading fertilizer in the flat.*

The Preplant Fertilizer Formula:

- 6 pounds di-ammonium phosphate (18-46-0), or treble superphosphate
- 4 pounds potassium—either sulfate or muriate of potash
- 4½ pounds ammonium nitrate or 7 pounds ammonium sulfate

4½   pounds magnesium sulfate (epsom salt)
4   *ounces* boron (borax) sodium borate or boric acid

Spread separately:

11   pounds lime—either agricultural, dolomite, or gypsum (see note on lime, page 41)
____
30   pounds total

Note: Wherever the preplant formula is pre-packaged and sold, omit the lime. Calcium (lime) is included in the mixture.

Also, when using pre-packaged preplant fertilizers, apply 15 *pounds* to each standard 5' × 30' grow-box (or one-half the quantity of one 30-pound bag).

Be aware that mixing the separate compounds together lowers their melting point in some cases. This happens because they are hygroscopic. After being mixed together the compound can become damp and wet within a few-days time, or it can set firm and hard. These changes do *not* weaken or affect the potency of the fertilizer formula, but it makes it inconvenient to apply. Because of this, it is recommended that the formulas be mixed in the amounts needed and applied the same day, if possible.

A bag of grow-box fertilizer.

Store both the separate compounds and the mixed fertilizers in a cool place at all times; never set them in full sun. Keep the bags closed, or provide containers with tight lids to store the various chemical compounds.

A few words about lime: *Use gypsum in areas getting less than 18 inches of rainfall annually. Use agricultural, dolomite, or slacked lime in areas getting more than 20 inches of rainfall annually.*

How Many Seeds Per Seedflat?

The size and the number of seedflats required depends on how many seeds will be planted. Frequently, several seedflats are planted at one time. Limit the seed to between 600 and 900 per flat.

Mixing soil and fertilizers.

Leveling the soil in a flat.

A flat filled and watered.

Making seed depressions with plastic pipe.

LET'S GROW TOMATOES

Step three: Fill the seedflats level full with special soil. Remove any excess soil by pulling a 1 × 4 × 20-inch straight-edge board over the top of the flat.

Step four: Gently and lightly water the seedflats to settle the loose soil.

Step five: From a ¾ or 1-inch piece of plastic pipe, cut off a length 16 inches long.

Step six: Make depressions in the seedflats by pressing the pipe into the soil evenly to the desired depth. The depressions can be made 2 inches apart across the seedflat. Not all seeds are planted the same depth. The size of the seed determines usually how deep the depressions should be made. For tomato seed, make the depressions in the seedflat ¼ inch to ½ inch deep.

Step seven: Scatter the tomato seed evenly along the bottom of the depressions in as *wide* a band as possible. Do not scatter more than 600 to 900 seeds in one standard-size 18″ × 3″-deep seedflat.

Step eight: Cover the seed by pulling a knife blade between the depressions, and gently flatten the surface level and even.

Spread the seed in a broad band.

Level the seeded flat.

Step nine: Water the flats sufficiently to settle all the loose soil, but *do not* float the seed to the surface. Use *only water* to water newly-planted seed. *Never* water unsprouted seed with fertilizer solutions! Why? Any kind of fertilizer applied to unsprouted seed can delay germination and can kill swollen seeds!

Step ten: Cover the planted seedflats with burlap or cheesecloth.

Use burlap to cover the seeds.

A can with holes is used for watering.

Step eleven: Keep the seedflats damp at all times. Whenever it is necessary to water, water over (through) the burlap cover. Remember, when watering germinating seeds in seedflats, *do not* roll or float the seeds to the surface.

Step twelve: Immediately after the new sprouts can be seen, remove the burlap and set the seedflats in full light. Do this even before the sprouts have emerged through the soil.

Step thirteen: And, just before removing the burlap, water the seedflats with the "Constant Feed" solution. Note that the sprouted seeds and seedling plants always should be watered with the "Constant Feed" solution during the entire time the plants are in the seedhouse. (The "Constant Feed" formula is given on page 48.)

7

Transplanting Seedlings

Newly-sprouted seedlings will require transplanting in 7 to 10 days after they have emerged through the soil. Transplanting seedlings into pots is recommended.

Square plastic pots are preferred over round pots, peat pots, or pots made of other materials. Remember, the special seedhouse boxes are called "flats." Flats sometimes vary in size and depth. But the recommended size is 18″ square by 3″ deep. They have bottoms but no lids.

Each flat holds 72 or 81 2-inch square plastic pots; 36 3-inch square pots; or 25 4-inch square pots.

Seeds germinated evenly.   *Transplanting seedlings.*

TRANSPLANTING SEEDLINGS

Filling Square Pots

The special grow-box soil outlined in previous chapters is used to fill the pots.

Here's a fast method to fill the pots with soil:

1. Fill the flats with square plastic pots placed side by side.
2. Put about two shovels full of the special soil over the pots in the flat.
3. Spread the soil over the pots by hand, and fill every pot with soil.
4. Remove all excess soil by pulling a 1 × 4 × 20-inch straight-edge board over the top of the filled pots.
5. After filling the pots with soil, water the flats gently to settle the loose soil particles.

Filling pots with soil.

Remove excess soil.

Planting Instructions

Step one: To make transplanting easy, take a ½-inch dowling rod and cut off a dibble 6 inches long, and point one end. Take the dibble and make a hole in the center of a pot. Dibble holes made in pots which are watered adequately *will not* cave in! The holes are clean. The hole should be the full depth of the pot. Select only the best-shaped and most vigorous plants for transplanting into pots.

Step two: Use the dibble to gently loosen the seedlings and lift a plant from the seedflat. Lift the seedlings *by the leaf, not* by the stem, and keep as much soil on the roots as possible.

Step three: Plant only one plant in each pot. The hole in the pot should be large enough and deep enough to accommodate the roots and stem of the seedling plant. Special care should be taken during transplanting to be sure the roots *do not* fold upward around the plant stem like a fishhook.

Make 6-inch dibbles from ½-inch dowling rods.

Wet soil leaves clean holes.

Step four: Transplant the seedlings *deep* in the pots. Leave only about ½ to 1 inch of the growing tip sticking above the soil surface.

Step five: Close the hole around the stem and roots by pushing the dibble, on an angle, down beside the plant stem. Be careful not to injure or bruise the plant stem in this process, but be sure the soil is in contact with the roots. Soil contact around the roots is essential for immediate growth of new roots. To check for soil contact, gently pull on the leaf. The plant should be tight.

Lift seedlings by their leaves.

Plunge the plants deep.

TRANSPLANTING SEEDLINGS

Step six: After planting, place the flats on a level surface such as a table. This is important to produce plants of uniform size.

Step seven: Water the planted pots very soon after transplanting, before the plants begin to wilt.

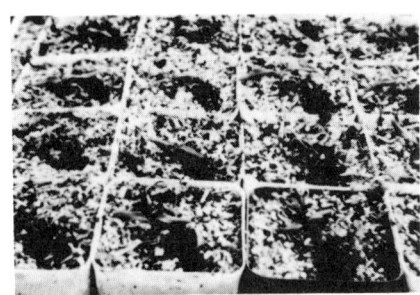
Water the plants before they wilt.

Don't bruise the plant stem.

Checking for loose plants.

The seedlings grow rapidly.

Remember, the modern concept in watering and feeding seedlings is the "Constant Feed" method. With this method plants are fed everytime they are watered. The "Constant Feed" method is usually continued throughout the entire time the plants are growing in the seedling greenhouse. The "Constant Feed" method is outlined in chapter 8.

8

Constant Feed Method

The Mittleider Weekly Feeding Formula

The "Constant Feed" fertilizer solution is made by using the following fertilizer formula:

<pre>
9   pounds calcium nitrate
4   pounds ammonium nitrate
1½   pounds di-ammonium phosphate (18-46-0)
4½   pounds potash (either sulfate or muriate of potash)
6   pounds magnesium sulfate (epsom salt)
8   ounces iron sulfate
4   grams copper sulfate
8   grams zinc sulfate
12   grams manganese sulfate
12   grams boron (sodium borate called borax)
3   grams molybdenum (either sodium or ammonium
        molybdate)
─────
25½   pounds total (strong)
</pre>

Make a "Constant Feed" solution.

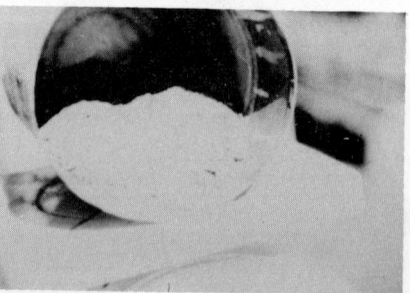

Weigh the fertilizers accurately.

This formula weighs 25½ pounds (strong). Whenever trace minerals (micro-nutrients) are used the quantities are very small. This is the reason they are called "trace minerals." And because of the small amounts, weighing and mixing together smaller quantities of this formula is not recommended.

Dissolve the fertilizers.

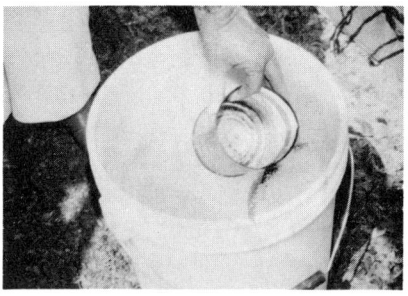

Can with holes in the bottom for watering.

One batch of this formula is adequate to supply the fertilizer needs for one standard 5' x 30' grow-box, 10 weeks minimum (13 feedings).

Procure a 25 or 50 gallon drum or similar container that will hold water. Plastic is recommended because it resists corrosion.

To make 50 gallons of "Constant Feed" solution, weigh accurately 1 pound (16 ounces) of the Mittleider Weekly Feeding fertilizers, mentioned above. And, for 25 gallons of solution, weigh 8 ounces of the same fertilizers. Fill the container with water and dissolve the fertilizers.

How To Apply The Solution

There are at least two methods to choose from in applying the "Constant Feed" to greenhouse plants.

Dip the solution from the drum into a sprinkler can and apply the contents to the plants with the sprinkler can.

Or, take a No. 2 can, cut out the lid from one end and with a small nail and hammer, or with an ice pick, make holes (many small holes) in the lid of the other end. Dip the solution from the drum and fill a 3 or 5 gallon pail. Dip the contents from the pail with the No. 2 perforated can and water the plants in the seedhouse.

Applying the "Constant Feed" solution to seedlings with the perforated No. 2 can is the quickest and most thorough method of application, but it may require a short period of practice before efficiency is attained.

Every time the plants are watered use this fertilizer solution, even if this is every day or twice a day!

When the drum is empty, refill it with water and add the fertilizers listed above.

The "Constant Feed" method will *not burn* the leaves of even the most salt-sensitive plants, and when using this method it is impossible to over-feed or over-water the plants, regardless of their size.

9

How To Keep Plants From Growing Spindly

Seed can be planted early enough to have plants 8 to 12 weeks old when the danger of frost is past. But, whether or not this can be accomplished depends on the propagating facilities available and if the seedlings can be cared for properly.

Under suitable but possibly less than ideal conditions, it takes between 5 and 8 weeks after seed germination for tomato plants to grow 8 to 10 inches tall. Much depends on the temperature. Tomatoes are heat-loving plants and grow very slowly when the soil temperature is 50° or less. In daylight temperatures between 80° and 90°, tomato plants grow rapidly.

"Constant Feed" produces healthy plants.

Properly fed plants remain healthy.

If the propagating facilities can supply young seedlings with adequate light, oxygen, nutrients, moisture, and warmth, the plants can be started early enough to have flowers and small tomatoes by the time the danger of frost is past and it is safe to plant them in the grow-boxes or garden.

To keep plants growing normally, provide sufficient nighttime heat to keep the temperature above 50° if possible, but at least above 32°. Plant growth is nearly stopped at 50°F.

During the daytime when the temperature is above 32°, allow fresh air into the seedhouse to change the stale air. During days when the outside temperatures are in the high 30's, opening the doors or ventilators for just 30 minutes in the seedhouse is adequate to supply fresh air to the plants. As the days lengthen and the outside temperatures rise, longer periods of ventilation should be given.

Water and feed the flats every day, if necessary. Never permit the soil to become dry!

Under normal conditions, plants overcome the transplanting shock within two or three days, and after 4 or 5 days the plants will begin to show new leaf growth.

Within two or three weeks after transplanting into 2-inch pots (or 3-inch plastic pots, depending mainly on the temperature), the leaves of the plants will begin to overlap each other in the flats. When this occurs tomato plants *will grow spindly* (leggy) and tall very quickly, since each plant is vying for adequate light.

Preventing Plants From Growing Spindly

When the leaves begin to overlap the leaves of other plants, the experienced grower who insists on thick-stem plants prunes off the leaves which overlap. Pruning off the leaves increases the light around the plant stems, and does not stop the growing tip from growing. Pruning off the leaves temporarily stops the upward growth of the plants, and encourages the stems to thicken, which is what the grower desires.

During this temporary period of reduced growth in the plant, the grower has two choices:

HOW TO KEEP PLANTS FROM GROWING SPINDLY

1. to wait 7 to 10 days for new leaves to grow and overlap again and pinch as before,
2. or, he can shift the plants into larger pots or gallon-size containers.

In the first choice the leaves must be pruned off again at the proper time to keep the plants from getting spindly, using the same procedure as explained earlier.

In the second choice, shifting the plants into larger containers provides more space between plants, delays pruning until a later date, and encourages the stems to thicken—due to increased light and circulation around the stems.

If, therefore, space is available in the seedhouse to accommodate larger containers, it is recommended that the plants be shifted from the smaller pots into 4-inch pots or gallon containers before they are pruned the second time.

Large, healthy leaf-growth.

Pruned plants.

10

Producing Large-Size Plants

Producing sturdy, healthy plants in gallon-size, or larger, containers is thrilling and rewarding. The process can lengthen the harvest season of vine-ripe tomatoes up to 12 weeks.

How To Shift Plants Into Larger-Size Containers

Step one: Fill the 4-inch pots or gallon size containers with the same special soil which was used to fill the smaller pots.

Step two: Water the soil until it is uniformly wet.

Step three: Use an empty flat or suitable container on the seedhouse table to hold the transplanting soil. A large plastic container is ideal to carry the soil for transplanting.

Step four: Before disturbing the roots of the plants to be transplanted, water the pots quite heavily. This is to keep the soil-ball from crumbling away from the roots when the plants are removed from the pots.

Fill the 4-inch pots with special soil.

Water with the "Constant Feed" solution.

PRODUCING LARGE-SIZE PLANTS

Step five: Now pick up a potted plant. Carefully place two fingers from the same hand on the topside of the pot—one finger on each side of the plant stem.

Step six: Without removing the fingers, turn the pot upside down and gently tap the top edge against the flat or table.

As the plant slips out of the pot, hold it gently and firmly with the two fingers still in position. Carefully turn the plant over and gently lower it into an empty gallon-size container.

If the growing tip of the plant *does not* stick up above the container at least one inch, drop soil into the bottom of the pot and gently lift the plant roots accordingly. The growing tip must never be covered with soil. It must have light to live.

Proper handling of plants when shifting to larger sizes will not crack or split the rootball and will set the plant deep.

Step seven: When the plant is set the proper depth, fill the container with soil and press it firmly around the plant stem with the thumbs or fingers. Add sufficient soil to fill the container within ½ inch from the top.

Step eight: After plants have been shifted, water them before they wilt. The water should contain the "Constant Feed" fertilizers, mentioned in chapter 8, and sufficient solution should be applied to saturate the soil in the container with enough extra that a little will seep out the drains in the containers. This first watering is always very important.

Step nine: Set the pots on a level surface, such as the tables in the seedhouse.

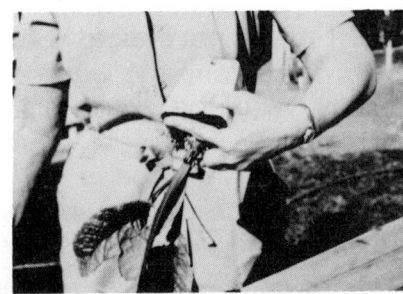
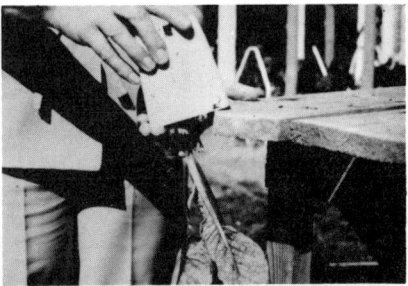

Proper position of fingers.   *Tap the pot on the edge of a table.*

LET'S GROW TOMATOES

Hold the plant firmly.

Lower the plant into a gallon pot.

Add extra soil and pack it lightly.

Proper soil level in the pot.

Daily care of plants in these gallon-size containers is the same as was outlined for smaller pots in a previous chapter.

From the time the *seed is planted* until the seedlings are transplanted into 4-inch pots is usually 3 to 4 weeks. All the time the seedlings are growing, if the plants are grown in the springtime, the days are getting longer and the weather is getting warmer. These changes make the seedhouse warmer and results in faster plant growth. Daily ventilation of the seedhouse should be correspondingly increased.

If the plants were graduated from 2-inch pots to 4-inch pots, their leaves will begin to overlap the leaves of other plants in about two weeks after the plants are shifted. When they reach this stage the leaves which overlap other plants should be pruned or cut off. This procedure was explained in a previous chapter.

PRODUCING LARGE-SIZE PLANTS

But use caution: *do not* prune off the young leaves near the growing tip, and *do not* prune off the growing tip, which is called the terminal bud!

If the plants have had adequate light they will be approximately 6 inches tall when this pruning occurs. As mentioned earlier, pruning off the leaves temporarily stops the upward growth of the plants and forces the stems to thicken.

Within 7 to 10 days after the first pruning in the 4-inch pots, the plants will need to be pinched again. The second pruning of the overlapping leaves can be quite severe. This time prune off all the leaves except the growing tip.

After the second pruning (pinching), the plants will have strong, thick stems and will be about 8 to 12 inches tall.

Do not pinch the plants more than two times in the 4-inch pots, or in larger containers. Further pinching produces undesireable woody stems! And be sure to allow *all* flower buds to develop wherever they appear on the stem.

If the plants must be kept in the seedhouse longer than two weeks after the second pinching in 4-inch pots, the plants should be graduated into gallon containers, or larger, within 2 or 3 days after the second pinching. Gallon-size containers are large enough to accommodate tomato plants 4 to 8 weeks without injury.

Transplanting plants into gallon containers was explained earlier. Review the steps outlined above. Planting into gallon containers is accomplished the same way as shifting plants into 4-inch pots. Review the steps outlined on pages 44-47.

Additional Information on Gallon-Size Containers

For the first 7 to 10 days after shifting plants into gallon pots, the containers can be lined up one against another on the level tables. But as before, when the leaves start overlapping the other plants in the pots, more light must be supplied around the plants to keep them from getting tall, weak, and thin stems.

This time light is supplied to the plants by gradually increasing the space between the containers. To increase the light factor the first time, separate the containers 2 inches apart on all sides. Thereafter, each time the plants require more light, increase the space between the containers accordingly. The aim in separating the containers is to provide enough light around the plants to produce thick stems and thrifty plants.

Plants can be grown in the larger containers 4 to 8 weeks before they must be transplanted into the grow-boxes where they will produce the crop.

Remember, the same special soil is used, whether it is to fill pots or larger containers. Also, the feeding solution and watering procedures are the same whether the plants are newly-sprouted seeds, growing in 4-inch pots, or growing in gallon-size or larger containers. And every time the plants in the seedhouse are watered, the "Constant Feed" solution is used!

Prune plants only two times.

Properly-pruned plants have strong stems.

Separate the cans to provide increased light.

Always use grow-box soils.

In summary, very explicit details have now been given on how to grow plants from seed, preparing the soil for transplanting, transplanting procedures, feeding and watering seedhouse plants, pruning young tomato plants and spacing containers to provide adequate light to each plant to keep them strong and vigorous.

11

Pruning, Staking, Tying

Producing container-size plants was thoroughly covered in the previous chapter. There are two special factors which are closely associated with growing large-size plants in containers which will be discussed in this chapter. These factors were omitted previously for fear of confusing the reader. The two factors are (1) pruning off suckers and (2) staking and tying grown plants in large-size containers.

Suckers and Pruning

Tomato plants produce suckers at all stages of growth. Above every leaf node (the place where a leaf is attached to the stem) is a bud which is called a "sucker." When the plants are small and the older leaves are pinched off the first or second time, the suckers are barely visible and are usually not noticed. After the plants are shifted to 4-inch pots, the suckers are usually large enough to be seen. In gallon containers, the plants and the suckers grow more rapidly than they previously did.

All suckers should be pruned from the plants as early as possible. This can be done either by rubbing them off by hand or by cutting them off with a knife. But remove only the suckers! *Do not* cut off the leaves or the flower buds.

PRUNING, STAKING, TYING   61

Use a knife to remove suckers.

A plant with suckers removed.

Tiny suckers are visible.

Cut off suckers as early as possible.

Staking and Tying

Sturdy, large-stem tomato plants will stand upright, without support, until they are 14 to 18 inches tall. Plants which were started from seed and grown as outlined in this publication will reach this height in 9 to 10 weeks after the seeds have *sprouted*. Therefore, plants in gallon containers should be staked and tied when they are about 9 weeks old.

This means also that plants which have been *shifted* to gallon containers from smaller pots will require staking and tying within 2 or 3 weeks after they are shifted.

Here's how to stake tomato plants:

Step one: From willow trees, dowling rods, etc., cut off stakes 18 to 28 inches long and about ½-inch thick.

Step two: Insert one stake along the side of the plant stem in the gallon containers.

Step three: Cut strings 10 inches long, one for each plant.

Step four: Before the plant starts to fall over, tie a string securely to the *stake first,* and then tie the plant to the stake. Leave the string loose enough around the plant to allow the stem to enlarge as it grows.

Put the stake beside the plant.

Tie the plant to the stake.

Stake the plants while they're small.

Staked plants in gallon pots.

Flower Set

Greenhouse tomato plants growing in cool temperatures will produce their first set of flower buds in approximately 4 to 5 weeks after the seed has sprouted.

If the plants have thick stems and are vigorous, the first set of flower buds will develop on the stem 8 to 12 inches above the soil level.

The height that the first flowers develop on the stem is governed by several factors such as fertilizing, light intensity, variety, growing temperature, drainage, soil oxygen, pruning, and depth of transplanting into pots and containers, etc.

Leaves

Vigorous plants have large, deep-green colored leaves. The leaves are the most important vegetative part of the plant. They affect the yield, the size, the shape, and the quality of the fruit.

12

Managing Single-Stem Vines

The vines of growing tomato plants in the grow-boxes should be kept up off the ground. This can be done in two ways: one is called the "stake" method, the other is the "A" frame method. Either method is satisfactory. The cost for materials is the same, and the number of plants and the space between the rows of plants in each grow-box is the same.

When the "A" frame method is used, the plants are spaced a uniform 7 inches apart in the rows across the grow-box.

When the "stake" method is used, the plants are planted in pairs 6 inches apart, with 10 inches between the pairs across the grow-box.

Staked tomato plants.

"A" frame structures.

The "Stake" Method

Transplant *8 plants across* the grow-box in each row as follows:

1. Transplant the first plant 2 inches *in* from the grow-box frame.
2. Plant the second plant 8 inches in from the grow-box frame.
3. Plant the third plant 18 inches in from the grow-box frame.
4. Plant the fourth plant 24 inches in from the grow-box frame.

Four plants fill *one-half* the row in the grow-box. To complete the other half of the rows, go to the opposite side of the grow-box and repeat the process.

In each planted row across the grow-box there will be four pairs of plants 6 inches apart, with 10 inches between the pairs.

Note that the plants are planted in pairs close together so that *two plants* can be tied to one stake.

Staked plants.

Installing Stakes

Step one: Drive a 2" × 2" × 8' stake between the pairs of plants using 4 stakes for each planted row.

A word of caution: Beware of substituting 1" × 2" × 8' stakes! When the vines and tomatoes are 6 and 7 feet high, even mild winds can break the 1 × 2-inch stakes very easily.

Drive the stakes through the grow-box soil down into the soil below, about 12 inches. The objective is to tie two tomato plants to each stake, and keep tying the vines as they grow taller until they reach the tops of the stakes.

Step two: Before tying the vines to the stakes, prune the plants to single-stem vines and remove the suckers. Use rot-resistant strings 18 inches long for tying the vines.

As often as needed, tie a string around the *stake first,* so it cannot slip down the stake later, and then tie the vine to the stake. Leave enough space between the string and the tomato stem to allow it to thicken as it grows.

Please note that the cultural methods recommended for growing tomatoes are the same whether using the "stake" method or the "A" frame method.

A simple way to drive stakes.

The "A" Frame Method

Properly-constructed "A" frames hold the crop up off the ground.

Here's how to make "A" frames: Use strips of 1" × 2" × 7' lumber for the upright stakes, and 1" × 2" × 5'-strips of lumber for across the grow-box.

Each completed "A" frame straddles one row of plants.

"A" frames are 7 feet high. They should be braced with angle braces and be strong enough to carry 150 pounds of leaves and fruit.

Braced "A" frames.

At the end of each row of plants and on both sides of the grow-box, nail a 1" × 2" × 7' strip onto the outside of the grow-box frame.

Next, across the top of the 7-foot strips, nail a 5-foot 1" × 2" strip in line with the row of plants in the grow-box.

Then, at the base of the 7-foot strips, beside each row of plants, nail a 1" × 2" × 5' strip across the grow-box.

In the "A" frame method, the tomato vines are supported by strings tied to the "A" frames. Materials recommended for tying the vines are twine, hemp, plastic, etc.

"A" frames and strings.

Beware of small-diameter size strands of fish line, nylon, etc. These are not recommended. They are too thin and can easily injure or girdle the stems of plants. The materials used for tying should have a tensile strength of 50 pounds per square foot. (Tensile strength refers to the load necessary to produce a rupture in a given material when pulled in the direction of its length, commonly expressed in pounds per square inch.)

How To Fasten The Strings To The "A" Frames

First, tie one end of the string to the 1" × 2" × 5' strips overhead, using one string above each plant. Guide it around the vine two or three times.

Next, thread the loose end of the string under the 1" × 2" × 5' strip, then bring it up over the strip and tie it securely. When properly tied, the strings should have a little bit of slack. They are not ridgedly tight.

"A" frames also can be constructed with strings tied in place before the plants require support. No matter which method is used, the plants should be tied to the strings before they begin to fall over.

Guide the vines around the strings.

Put "A" frames in place early.

13

Spacing Tomato Plants In Grow-Boxes

In the past, field-crop tomatoes, which were grown for marketing, were planted 4 feet apart, each way, one plant per hill. The plants grew vigorously and the leaves and vines covered the ground between the plants. The yields generally were heavy and the quality satisfactory.

Today, because of the high cost of hand labor, production cost, diseases in the soil, insects, and other factors, market-grown tomatoes are produced quite differently. For example, the vines are usually kept up off the ground by tying them to wooden stakes or strings, the plants are pruned to single stems, and the plants are grown close together. The grow-box method simplifies all of these procedures.

The standard size grow-box is 5 feet wide by 30 feet long by 8 inches deep (outside dimensions). The aisles are 3 feet wide between the boxes and 5 feet wide between the ends of the boxes.

Tomatoes are planted *across* the width of the boxes, not along the length, and the space between the rows of plants is a minimum of 28 inches.

The spacing between the plants in the row is 7 inches, and there are usually 8 plants per row. The first row of plants should be planted 4 inches in from the end of the grow-box. The first plant in each row should always be planted as close to the grow-box frame as possible (as close as 2 inches). This increases the amount of space between the plants in the rows, and thus increases the very important light factor around each plant.

Calculating The Number Of Plants One Grow-Box Holds

Following the planting recommendations given above, a 30-foot grow-box will accommodate 14 rows of plants. Fourteen rows by eight plants per row equals 112 plants in one grow-box.

Usually 8 to 16 tomato plants will provide adequate vine-ripe tomatoes for table use for a family of 6 to 8 people.

In grow-boxes you can plant whatever number of tomato plants needed. The important consideration when planting less than a full grow-box with tomatoes is to plant so they will not shade out short-growing crops. Tomato vines will reach 7 feet high in 6 to 8 weeks of growth.

What Size Plants Should Be Transplanted?

The ideal plant size for fast and convenient planting to grow-boxes or fields is 8 to 12 inches tall. Gallon-size plants are an exception, and the planting procedures for large container-grown plants will be explained in a future chapter.

Individual tastes vary, and it is unlikely that two equal-sized families will favor either the same varieties or the same quantities of fresh vegetables. Also, no two growers harvest the same yield from the same crops. For example, some produce cabbages averaging 4 to 5 pounds per head, while for another grower the size will likely average only two pounds per head.

Tips On Mixing Crops In Grow-Boxes

Construct grow-boxes to face east and west, or any direction except north and south, if possible. Rows of plants are planted *across* the grow-boxes, not lengthwise. If the boxes face north and south the shadows from the crops change very little, if any, during daylight hours. Whereas, if the boxes face east and west, the shadows from the plants change all through the day. Light is very important to both yield and quality and also in the control of disease, such as powdery mildew, botrytis mold, etc.

SPACING TOMATO PLANTS IN GROW-BOXES

Stakes keep the crop off the ground.

When planting grow-boxes, always plant the crops which will grow the tallest to the north of shorter-growing crops. The sun's position in the sky changes from near center to the south week by week as the season changes from summer to autumn. Therefore the shadows fall to the north. Planting short crops to the north of taller crops robs them of sunlight and can result in crop failure.

Still another consideration when planting grow-boxes is in the number of days crops take to mature. As far as practical plant fast-maturing crops to the south end of the grow-boxes. Some crops mature fast enough to permit growing a second and third crop the same season. If such crops are planted on the south end of the grow-box, as quickly as the crop is harvested another crop can be planted in the same plot without delay or interfering with other crops.

Grow-boxes are versatile! A 5' x 30' box can easily accommodate 8 or more varieties of crops at one planting, but experience will indicate that a maximum of 4 varieties growing together in one grow-box is a better practice.

A good reference book.

14

Transplanting Procedures Illustrated

Step one: Generously water the grow-box soil.

Step two: Take a marker and mark off the rows along the length of the grow-box, and also mark the rows across the width of the grow-box.

Step three: Assuming that the plants are 10 to 12 inches tall, make a hole by a mark, large enough around for the plant roots and about 8 inches deep.

Step four: Lower the plant in the hole. Raise or lower the plant as necessary to leave only about 3 inches of the growing tip sticking above the soil level.

Step five: Then, with one gentle forward movement of the hand, fill the hole around the plant stem with soil.

Step six: Press the soil gently and evenly around the plant.

Step seven: Water the grow-box gently and adequately to settle the loose soil firmly around the stems and roots.

"Custom-made" soil absorbs water quickly.

Mark the length first.

LET'S GROW TOMATOES

Mark the width.

Make an adequate-size hole for the roots.

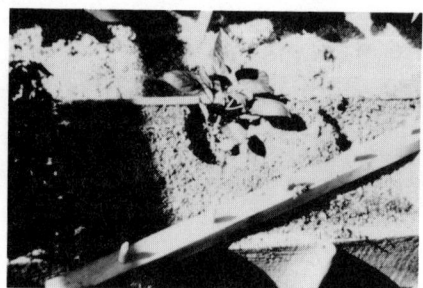

Set the plants deep.

Cover the hole and level the soil.

For more illustrations and photos on transplanting procedures, refer to the book *Food For Everyone,* mentioned previously, page 434, Figures 51:69 through 51:71.

In a previous chapter, the statement was made that plants 8 to 12 inches tall are considered the ideal size for fast and enjoyable transplanting into grow-boxes or gardens. This raises the question "If plants are 14 to 18 inches tall, and are healthy, can they be used?" If they are transplanted properly the answer is *yes!*

How To Plant Extra-Long Tomato Plants

Mark the grow-box the same as for other plants.

By a mark, make a hole 8 inches deep and a horizontal trench 8 to 12 inches long, depending on the length of the plant, parallel with the hole.

Take a long plant. Lay the roots and stem in the trench and carefully bend (but do not break) the growing tip up. Leave the growing tip sticking out above the soil surface about 4 inches.

Cover the roots and stem in the trench and gently pack the soil.

Level the soil around the plant and water moderately heavy. After a few days, new roots will grow out all along the buried stem.

With the growing tip exposed to light and air, the plant will develop sturdy growth. If the plants are fed and cared for adequately, their performance will equal that of any well-grown plants.

Dig the hole and trench.

Put the long plant in the trench.

Lift up the growing tip.

Cover the trench and hole.

During transplanting, always set the plants deep, whether in the field or in the grow-boxes. When possible, leave only 2 to 4 inches of the growing tip sticking above the soil surface. (It is equally as important for other vegetable crops to be transplanted deep! This characteristic feature distinguishes the difference between the amateur grower and the expert.) Planting deep enough to cover the growing tip (crown), however, will kill the plant in 3 or 4 days.

After transplanting, the plants should be fertilized and then watered. Use the Mittleider Weekly Feeding Formula given on page 48.

15

Pruning Fruiting Tomatoes

Pruning young tomato plants growing in pots and gallon-size containers was explained in chapter 11. In this chapter, pruning plants which are bearing fruit is explained.

The art of pruning tomato plants is best mastered through practical experience. It is something like swimming—you can study all about it, but learning to swim comes with getting in the water.

Prune tomato plants early.

The question which is frequently asked is *why prune?* The main reason is to increase the number and the size of well-shaped fruit. But there are also other reasons. Crowding too many plants growing at random together will produce tall spindly plants with thin stems and small fruit and leaves. Such plants are disappointing. However, when tomato plants are pruned to single-stem plants, they can be grown close together without sacrificing either yield or quality. The key is prompt and accurate pruning! For pruning to be most effective it must be put into practice early, while the plants are small.

The first pruning for established plants should be for the gradual removal of the lowest leaves which touch the ground and all suckers, as early as practical. Moderate pruning is usually continued until the plants are 24 inches tall.

Suckers are missing.

Do not remove any flower buds on the main stem. Continue the gradual pruning of the lower leaves until there is about 12 inches of clearance between the leaves and the soil. The reason for this pruning is primarily as a precaution for disease control. Keeping the stems of plants dry and exposed to light and air at the soil surface is an effective way to reduce fungus disease epidemics.

The leaves are *very* important to plant growth and fruit development. Severe pruning should be avoided. Seldom should more than one full leaf-spike be removed from a plant in a single pruning.

To accomplish the pruning effect mentioned above, the operation should spread over several weeks and several prunings. And, after the initial pruning of the lower leaves has been accomplished, further *heavy* leaf pruning should stop. The only leaves that should be removed thereafter are those which have served their usefulness and show by their appearance that they are not contributing to the growth of the plant.

Caution When Diagnosing Symptoms

It is not unusual for the oldest leaves to turn a lighter green color than the youngest leaves and the leaf edges to roll upward. If this occurs only on the oldest leaves, they can be removed. But there are other factors that affect the color and appearance of the leaves besides age. And if the youngest leaves become discolored, chlorotic or nechrotic, it is fair warning that something is wrong. A careful diagnosis should be made and the proper treatment administered.

Hands are spaced 4 to 6 inches apart.

On vigorous plants, the *second cluster* of flowers is usually visible before the first flower cluster has finished flowering. The distance between the flower clusters on the stem depends largely on the variety and the light factor. For many main-crop varieties, the space between the flower clusters is from 6 to 10 inches.

Special care should be given to protect the growing tip on the main stem. If the growing tip is injured or broken off, the main stem stops growing at the point of injury. This means a decrease in yield.

Protect the stem from injury.

More Information On Leaves and Suckers

The leaves of tomato plants are attached to the stem by petioles, and are from 3 to 6 inches apart along the stem.

Just above every leaf node—the point on the stem where the petioles are attached to the stem—are buds called "suckers." Usually there is just one sucker per leaf-node. As the suckers grow, and while still very small, they should be removed. This can be done by rubbing them off with the fingers or by cutting them off with a knife.

Be sure to remove only the suckers. Save the leaves and the flowers.

In addition to removing the suckers, all young shoots which occasionally grow up around the stems of the plants at the soil surface should be removed. These shoots are true suckers. They come from buds growing on the stem below the soil surface.

After the first heavy pruning at the base of the tomato plants, the pruning procedure thereafter is primarily to remove all suckers which grow out above the nodes along the

A knife makes suckering easy.

main stem, plus any new shoots (suckers) which grow from the main stem at the soil surface.

Pruning tomato vines to single-stem plants is becoming more common year after year. Pruning and removing of suckers continues week after week until the main stem reaches the overhead 1" × 2" × 5' support, or the top of the stake, which is usually seven feet above the soil surface.

When the main stem reaches the top support or top of the stake, the growing tip should be cut off. This operation stops the stem from growing longer and diverts the energy from expanding the plant to ripening the tomato crop.

Tomato growers quickly learn that the price the crop brings is governed by the dates the crop matures. Therefore, growing schedules are important.

Single-stem plants are preferred.

PRUNING FRUITING TOMATOES

For main-crop varieties, it takes about 8 weeks from the time the *seed* is planted until the *first set* of flowers are pollinated. Then it takes about 8 weeks more to produce vine-ripe tomatoes after the flowers are pollinated.

Of course, the number of days required to produce ripe tomatoes from seeds depends largely on the temperature in the greenhouse or garden, and also on the variety of tomatoes planted. Thus, the actual total number of days may vary.

Even though large (one to three-gallon size) plants can be transplanted successfully, the fact still remains that the earlier, younger, and smaller the plants are when transplanted where they will produce the crop the better. Personal attention is minimized and there is less risk from losses due to transplanting shock and root pruning.

Single-stem plants are planted close together.

Staked plants are pruned.

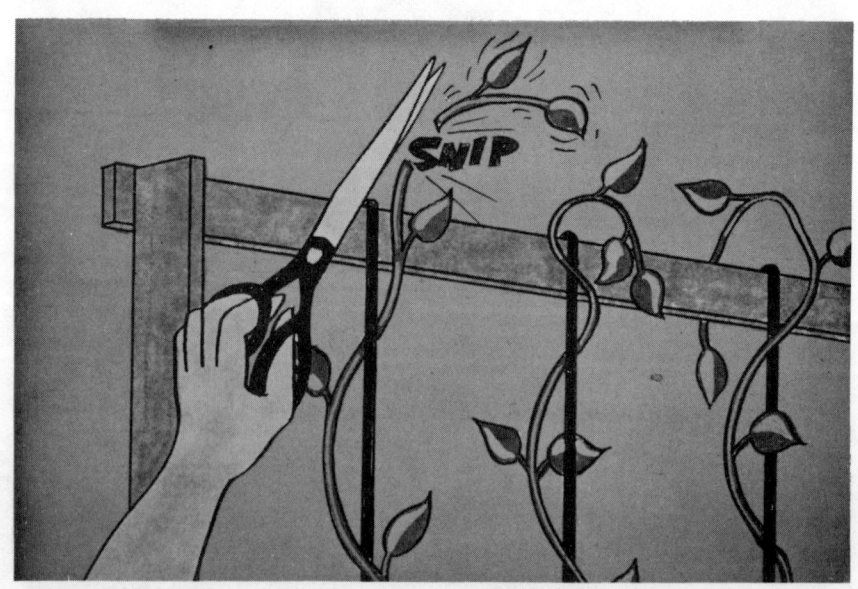

The growing tips are cut off.

16

Daily And Weekly Care

"A" frame construction was explained in chapter 12.

When strings are used, their purpose is to support the vines as they grow longer and taller.

Before the plants begin to bend over, the tomato vines are guided (not twisted) around the strings. One string is provided for each plant. The string is pulled over the side of the grow-box, then threaded under the wire and tied securely to the wire.

The process of guiding the vines around their respective strings goes on at least *once every week,* as the terminal buds grow until they reach the 1" × 2" strip or overhead wire where the strings are tied.

When they reach the wire, the growing tip is cut or broken off and the stem stops growing. Along with guiding the vines

Provide one string per plant.

Bring the string over the side of the box.

around the strings, the suckers and all broken or unhealthy leaves are removed. Note that the vines should be guided, never twisted, around the strings, and no decaying leaves or plant parts should be allowed to remain among the tomato plants or in the aisles or beds of the growing area. A healthy tomato crop requires a clean growing environment.

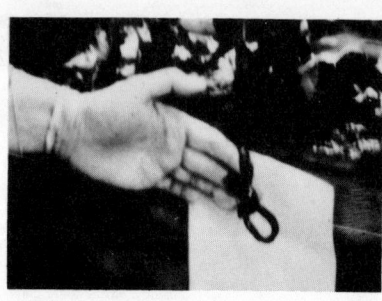

Tie the string around the wire.

Plants are cut off when they reach the top.

Break off the growing tip.

Remove the suckers.

The figure of 7 feet has been used many times and no explanation has been given. Seven feet high is not a binding figure. But experience has shown that to perform the various functions associated with growing and harvesting a crop of tomatoes, this height is about the maximum limit for most gardeners. It is the gardener's reach, not the plant's size, that determines the optimal height.

Planting, pruning, and tying are important procedures in successful tomato production.

17

Watering Tomatoes

Almost 95% of a plant's weight is water. From the tip of the deepest root in the ground to the tip of the highest leaf in the air, a plant is one continuous water pipe.

Ironically, nearly 95% of the water which plants require is lost into the air through their leaves. This is called "transpiration." Through transpiration, plants keep cool on hot summer days. Obviously, tomato plants require a continuous supply of water. And therefore raises the obvious question, "How much water do tomatoes require?"

Plants are a water pipe.

The complete answer to this question involves another question of equal importance which involves "soil air." While giving answers, it is important to emphasize that "soil air" should be considered along with water requirements. Keep this thought in mind as the question above is answered through the following illustration:

Step one: Fill a gallon container with special grow-box soil—a mixture of sawdust, sand, and/or perlite. Be sure there are holes for drainage along the bottom edge of the container. Set the container on a table or bench.

Step two: Pour 1½ pints of water over the soil in the container. Wait 5 minutes and then examine the drain holes to see if water is dripping through. If the soil in the container was *dry,* the 1½ pints of water may not wet the soil beyond its holding capacity and no water will drain off. If this is the case, pour one more pint of water on the soil in the container. Wait 5 minutes more. This time water will be dripping from the drain holes.

The point being illustrated is that soils hold only specific amounts of water. They are *not dams* to store water for future use.

When drainage is adequate, soil oxygen is adequate, and good soils absorb water until they reach saturation—called "field capacity." When "field capacity" is reached, if another drop of water is added to the soil, a drop will drain off. Whatever amount of water is added after the soil has reached "field capacity," that amount will drain from the soil, if drainage allows.

The special soil in the grow-boxes takes water quickly and evenly. And through experience, each grower can learn how much water is required to water thoroughly and yet not excessively.

The Goal In Watering

The goal is to strive to apply enough *extra* water at each watering to force some water (very little) out the bottom somewhere along the sides of the grow-box.

WATERING TOMATOES

Water grow-box tomatoes daily.

The frequency of water applications depends on the weather, the variety, and the size of the leaf coverage; also the load of fruit the plants are carrying and the salinity (quality) of the water.

It is nearly impossible to overwater crops in grow-boxes which have been constructed properly and filled with the soils recommended in this publication.

With this in mind, it is highly recommended that tomatoes, or other vegetable crops in grow-boxes, be watered *adequately* every day throughout their growing and producing season!

Grow-boxes which are watered, as outlined above, will produce high yields and choice crisp, juicy, highly-flavored fruit, and high quality vegetables—providing, of course, that the other essential factors are satisfactory. That's the payoff!

Depending on the amount of water pressure and volume, it can take from 2 to 10 minutes to water one standard 5' × 30' × 8" grow-box adequately.

Water Requirements For Tomatoes

Comparing tomatoes with corn, tomatoes are slow to wilt when they need water. This may be the reason why they are sometimes neglected. Nevertheless, the fact remains they *require* daily applications to sustain a continuous supply of available water to all parts of the plant.

To assume that tomatoes will produce a bountiful crop without frequent applications of water is equally as serious as it would be for a poultryman to assume that the way to make hens lay more eggs is to cut short their water supply.

Tomato plants produce a heavy and extensive root system, and unless the drainage under the grow-boxes is poor or there is a solid layer of rock or clay which the roots cannot penetrate, the roots will penetrate the soil below the grow-boxes 12 to 18 feet. The roots of bush beans are just as extensive. Still they are watered frequently.

All things considered, the heavier and healthier the leaf area, the larger the yield will be, and the more water the crop will require for both transpiration and for its own use. Therefore, by the time tomatoes wilt, due to lack of water, considerable damage has already occurred to the crop.

Adequate care is rewarding.

18

Fertilizers

Every phase of crop production emphasizes anew that guessing is costly and should be eliminated as far as possible. This is especially true when fertilizers are used.

One of the important reasons for recommending the special grow-box soils is because they are very low in fertility. This makes it necessary to supply *all* the essential nutrients which plants must have, and makes it easier to fertilize accurately.

Making special soil.

When animal manures or compost are used to feed plants, the grower is largely guessing on the amounts of fertilizer the plants have access to. If trouble in plant growth develops, the cause is uncertain because the grower does not know whether the phosphorus or some other nutrient is high or low and therefore the crops can fail.

On the other hand, because the grow-box soil has virtually no fertility and the nutrient requirements of the crop are established, it is possible to fertilize accurately for every crop and with every application.

Since this is possible, every crop can be a success. And if trouble in deficiencies should develop, a diagnosis is possible, and corrections can be made with reasonable accuracy. All this is possible because the grower is dealing with known facts—he need not guess!

Feeding Established Plants

Feeding sprouting seeds and transplanted plants in pots and gallon containers was explained in chapter 11. In this chapter, feeding established plants either liquid or granulated fertilizers will be discussed.

It is important to keep in mind that ever-bearing crops such as tomatoes, cucumbers, pole beans, zucchini squash, melons, etc., require the essential fertilizers over a longer period of time than do the single-crop varieties like cabbage, head lettuce, beets, carrots, etc.

Of the ever-bearing vegetables, tomatoes very possibly produce the heaviest crops. It is important, therefore, to remember that tomato plants, while they are increasing in size, setting fruit and ripening fruit all at the same time, require plenty of fertilizer and water.

If the stems are thick to within 12 inches of the growing tip, and the leaves are a healthy dark green color, and the flowers are pollinating and setting fruit, the chances are good that the plants are getting sufficient fertilizers, at least temporarily.

If some essential nutrients are missing, or the fertilizers are not properly balanced, the plants will indicate that this is the case by the symptoms on their leaves and/or fruit. Thus,

FERTILIZERS

when trying to determine the fertilizer level in a crop, the following questions should be considered:

1. Are *all* the plants growing evenly and satisfactorily fast?
2. When the flowers bloom, are they large or small?
3. Do the flowers pollinate automatically, without man's help, and do the small tomatoes grow, or do they fall off?
4. Do the leaves—both the leaflets and large leaves—have a uniform living-green color without any scorch or yellowing on the edges?
5. Are the leaves large or small?
6. Are the small tomatoes uniformly shaped?
7. What percent of the maturing fruit develops cracks?

The information contained in these questions reveals the way the plants are growing.

Fortunately for the grower, if problems arise, the solution for making adjustments on fertilizers have been established. They are given in *Appendix 1: Nutrient Deficiencies.*

Healthy plants, healthy leaves.

Some of the solutions have taken considerable research to establish. And the grower can choose to accept the information that is available today, or he can experiment on his own. The risks in the latter choice are high and expensive.

Years ago someone made the following statement:

"Five dollars worth of borrowed brains can save you five-thousand dollars worth of borrowed trouble."

 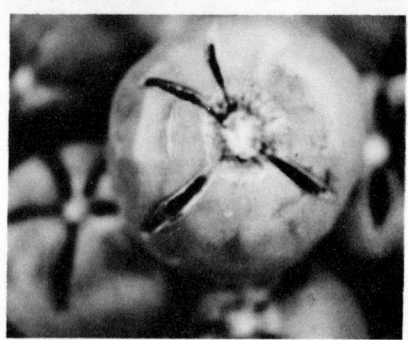

Hungry plants speak out. *Cracked fruit—a symptom of disorder.*

The Weekly Feeding Procedure

A satisfactory procedure for growing tomatoes in grow-boxes is to feed *once* each week. (This statement applies to fertilizing the standard-size grow-box *only*. Tomatoes grown in other size grow-boxes may require feeding more often!)

In a previous chapter the statement was made that tomatoes are ever-bearing. Therefore, before any fruit is ripe the vines are supporting considerable fruit and leaves, in various stages of maturity. Obviously, as the fruit load increases, the amounts of fertilizers should be increased at least once (and frequently twice) during the growing and producing season. Everyone understands that the appetite of a "mother-to-be" increases month by month until the baby is born. Similarily, tomatoes require more fertilizers as the load of flowers, leaves, and fruit increases.

Fine quality vegetables are produced when all the essential nutrients are adequate and in proper balance. A deficiency in *even one* essential trace mineral, such as boron, can result in crop failure.

Wherever crops are grown today, regardless of the soil they are grown in, fertilizers must be supplied if the crops are to produce. And it rests with the grower to supply the fertilizers in the right amounts and at the proper time.

From this discussion, it might appear that each plant requires a special combination of fertilizers and also different amounts. Fortunately, this is not the case!

All common vegetable crops can be grown on the same balanced mixture of fertilizers. And if the fertilizer mixture does contain all the essential nutrients, tomatoes will grow on the same formula used to grow other vegetable crops, and visa versa.

Two complete fertilizer formulas.

Two fertilizer formulas are given in this book: The "Preplant Fertilizer" formula is given in chapter 6, and the "Weekly Feeding Fertilizer" formula is given in chapter 8.

These formulas are recommended for feeding all varieties of garden and field crops, and are especially recommended for tomatoes.

Because these formulas have already been given, the only additional information needed in this chapter will be how much, how often, and when to increase or decrease the amount of fertilizers. Please remember, the feeding instructions in this chapter are *specific* for tomatoes, and do not apply to other crops.

Before proceeding with actual fertilizing instructions, it is assumed that the grow-box contains the special soil and the preplant fertilizers, that the soil has been mixed and watered properly, and that the transplanting of tomato plants in the grow-boxes is under way.

Be accurate as you measure fertilizer.

Actual Fertilizing Procedures

1. The *same day* the plants are transplanted, each grow-box should be fertilized accurately with the following amount of *straight nitrogen* fertilizer; either 1½ pounds of urea or 2 pounds of ammonium nitrate, *but not both.*

Spread the dry granular fertilizer beside each row of plants, at a distance of 4 inches from the plants.

FERTILIZERS

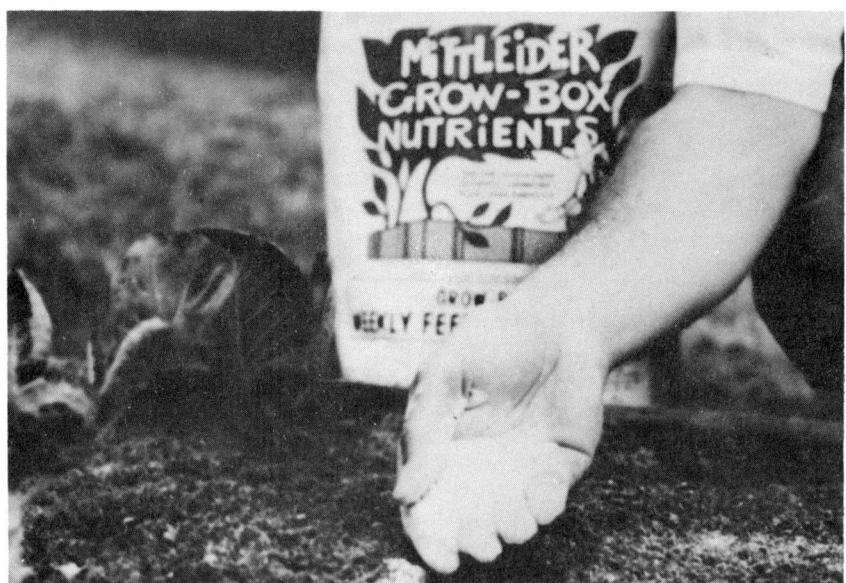

Spreading fertilizer.

A word of caution: *do not* place the fertilizers closer to the plants than 4 inches because fertilizers are concentrated mineral salts and, like common salt, will burn the leaves, stems and roots of plants, if placed directly on them before the fertilizers are dissolved and diluted with water.

Gently, yet heavily, water the grow-box between the rows of plants to dissolve the fertilizers. Be aware that a residue in some fertilizer compounds may not dissolve. These are not actual fertilizers but are the carrying agents, such as talcum powder, dimotacious earth, or similar materials, used in the manufacturing of the fertilizer compounds. These residues are *non-toxic* and *do not* interfere with plant functions or growth. They mix easily with the soil when it is mixed for another crop.

2. *Two days after* the first feeding, apply fertilizers again. Feed each grow-box the *same amount of nitrogen* fertilizer as applied the first feeding.

After applying the nitrogen fertilizer, water gently but heavily enough to completely dissolve the fertilizers.

Note that neither urea nor ammonium nitrate fertilizers leave any undissolved residue—all is water soluble.

3. *Three days after* the second feeding, feed the plants again.

Starting with this third feeding and onward, use *only* the *Mittleider Weekly Feeding* fertilizer mixture. Spread *2 pounds* per grow-box. Apply the fertilizer as described above. (The "Mittleider Weekly Feeding Fertilizer" formula is given in chapter 8 and Appendix 1.)

Water sufficiently to dissolve the fertilizer and carry it to the roots of the plants.

4. Feed the *fourth time nine days after* the first feeding. The amount of fertilizer is 2 pounds. Water sufficiently to dissolve the fertilizers.

A Review of Feeding Instructions

Newly-transplanted tomato plants in the standard-size grow-boxes require *2 feedings* with *only nitrogen* fertilizer compounds, and *2 feedings* with the "Mittleider Weekly Feeding" formula, during the *first 9 days* after transplanting.

Accuracy in this respect will result in the plants having a fast recovery from transplanting shock and rapid growth with the coveted living-green colored leaves.

After the first *9 days,* feed the plants *once every week.*

Each application should be *2 pounds per grow-box* of the "Mittleider Weekly Feeding" formula until the plants are 24 to 30 inches tall, or showing their third set of flowers, and then *2¼ pounds* of the same fertilizers *once* every week thereafter.

Please remember the plants are fertilized accurately *once every week,* but they are watered uniformly-heavy *every day,* six days per week. Regular daily watering is essential, especially in dry weather.

When To Decrease The Fertilizers

After 50% of the crop has been harvested, *reduce* the amount of fertilizer to *2 pounds* per grow-box every week.

Stop all fertilizing *three weeks* before harvesting is completed!

Note: The application rate and the feeding procedures are the same regardless of the kind of crop grown. If only a portion of the grow-box is planted, reduce the application rate accordingly. For example, if one-fourth of the box is planted, divide 2 pounds (the application rate) by 4, which equals ½ pound (8 ounces). The weekly feedings, in this case, would be 8 ounces, regardless of the kind of crop.

19

Flower And Fruit Set

Garden shops display an assortment of products which are sold for the sole purpose of keeping tomato flowers from falling off after pollination and to help set the fruit.

Obviously, it must be common for tomatoes to lose their flowers and their fruit, judging from the many products sold to correct this problem.

That this problem is not experienced by all growers, however, is evident from the literature available explaining tomato thinning techniques.

This much is sure! When tomato plants drop their flowers or fail to set fruit something is wrong! Quite likely, too, there are several factors involved rather than one.

Some of the more obvious factors which affect plant growth, and flower and fruit set, will be considered here.

Avoid shade.

Select a sunny location.

The Light Factor

Shade on the grow-box area from hedges and buildings, trees, north slope, etc., or from too many plants crowded together in the grow-boxes, will produce small leaves and thin stems. Such plants will be unproductive.

The solution to overcome this condition is to re-locate the grow-boxes where light is adequate all during the day, or thin out the extra plants, if too many plants are causing the problem. Plants, like people, must have living space!

The Heat Factor

The statement was made earlier that tomatoes are heat-loving plants. This is true! But there is a limit.

Temperatures between 75° and 95° are ideal for fast tomato production. Tomatoes will produce below 75° but their growth is slower and is further reduced as the temperature drops. Growth is nearly dormant at 50°F.

Also, tomatoes will grow in temperatures above 95° to about 100° without difficulty. But, every degree above 100° increases the problems of production. And during long days and temperatures above 115°, the pulp in the tomato fruit turns to liquid and the crop is not saleable.

This condition occurs with tomato plants growing in full sun without artificial protection. By diffusing the sun through shading, and by keeping the soil and aisles damp, tomatoes can be grown in areas where temperatures reach 120°F.

During periods of high heat, tomato flowers may fail to pollinate and set fruit. And it may be necessary to provide enough shading over the crop to lower the temperature to correct this condition.

The leaves of tomato plants should *not* wilt during the heat of the day!

The Plant Population

How many plants should one acre support? This question is frequently asked! The answer depends mainly on what the crop will be used for—whether it is intended for cannery, roadside fruit stands, or for greenhouse production!

Cannery Tomatoes

In areas where hundreds of acres of tomatoes are grown for canneries, it is common practice to plant the seed directly in the field, so transplanting is eliminated.

Some farmers continue to thin their crops, at least partially, while the majority have discontinued thinning altogether. According to the published reports on tomato yields grown for canneries, thinning is not justified. Cannery tomatoes are usually machine-harvested and picked only once.

The point to be made about tomatoes grown for canneries is that, regardless of the plant population in the rows, the plants hold their blossoms, pollinate, set and hold their fruit. Just one look at a tomato field after the mechanical pickers have gone through is enough to establish this point.

Market and Fruit Stands

With the exception of hydroponic production, the tomatoes which are grown for marketing and roadside fruit stands were formerly grown in the field without specialized care.

In recent years this has changed. Now more and more tomatoes are pruned and tied to stakes in the field. Plants which are pruned and staked can be planted much closer together in the rows, thereby increasing the acre yield and also the quality of the fruit.

The increased yield and the superior quality of pruned staked tomatoes compensate for the extra expenses incurred.

An average plant population for field-staked tomatoes is 6,000 to 8,000 plants per acre. The rows are spaced 60 inches apart and plants are spaced an average of 10 to 12 inches apart in the rows. Farmers which produce tomatoes for the market and fruit stands seldom complain that the flowers do not pollinate or fail to set fruit.

Greenhouse Tomato Crops

Greenhouse tomato plants are pruned to single-stem vines and guided around strings which are tied to overhead wires. Sometimes special-made plastic clips are used to hold the stem to the strings. Either method is satisfactory.

The vines are kept off the greenhouse floor and usually grow to a height of 7 feet before the growing tip is cut off.

A high yield crop.

The space between plants in the rows varies between 4 inches and 22 inches. The space between the rows varies between 48 inches and 56 inches.

Greenhouse tomato plants are pruned carefully. They receive excellent care when compared with plants growing in the field or garden. Frequently, however, greenhouse tomato growers complain about the flowers falling off, poor pollination, and disappointing fruit set.

Some growers give one reason why this happens and some give another. The trouble with giving reasons is they seldom solve the problems!

The Family Garden

Soils and planting methods vary widely, but the following complaint prevails. "Why do the flowers and tiny tomatoes fall off of healthy vines?"

Again, many answers are given but the problem remains. And the finger logically is pointed to something in the growing procedures.

The following several chapters will be a discussion on several possible factors which may be responsible, either singly or collectively.

20

Nematodes

Root Nematodes (Eel-like Worms)

All around the world the soils are infested with nematodes. Fortunately, there are methods available to successfully eliminate them. As was mentioned in the heading, nematodes are tiny eel-like worms. So tiny, in fact, that they are seldom seen with the naked eye.

These eel-like worms eat their way into the roots of plants and thereafter live off the essential liquids in the roots.

Nematodes.

Nematodes are especially fond of tomato plants. Infested roots have irregular brown-colored swellings which appear like rough knots.

Nematodes multiply rapidly and, as they increase, the knots become larger. One female can lay 3,000 eggs in a normal life-cycle.

Tomato roots can be heavily infested with nematodes without affecting the green color of the leaves or noticeably retarding plant growth. The visual appearance of infested plants can be normal, but plants which fail to set fruit are telling the grower something is wrong!

How To Inspect For Nematode Infestation

The real function of plants is to perpetuate their kind. The natural process for tomato plants to accomplish this is to produce seed. The seed is in the ripe fruit. Therefore, plants concentrate on producing flowers and fruit. When the flowers mature, they are receptive to pollination. The receptive period for a flower is about 6 hours. After the flowers are pollinated, which is generally automatic for tomatoes, the ovules (the tiny tomatoes at the base of the flowers) conceive and the ovule begins growing.

Conception places a heavy load on the plant. If the roots are functioning properly, and the essential nutrients, water, air and temperature are satisfactory, the roots can easily support the added responsibility, which is to develop the tomatoes. But if nematodes are living in the roots, plant performance is reduced according to the number of nematodes present.

How Plants React to Nematodes

By the time the plant is flowering, the nematodes have multiplied so much that the plant has a full load just to support the nematodes.

The plant cannot expel the nematodes; therefore, when the flowers mature and are pollinated, they are aborted along with the small fruit in order for the plant to survive and stay alive. But the struggling plant does not give up. It tries to produce seed. And it puts out new leaves, buds, and flowers.

And again, at the crucial moment, because of the nematodes, the plant aborts the fruit just to stay alive. This process is repeated again and again in nematode-infested plants. The plants cannot do otherwise.

Therefore, plants which appear to be healthy but which fail to set fruit should be carefully inspected for nematode infestation.

21

Fertilizer And Soil Problems

It is usually the case that garden soils become very hard and depleted of the essential nutrients during the growing season.

Average soils are hard and depleted.

FERTILIZER AND SOIL PROBLEMS

Only a very few people living in rural areas still have access to animal manures to improve soil fertility. The vast majority of the world's population must depend almost completely on lime and chemical fertilizers to feed their garden and field crops. This condition will not change. Therefore, it is important to learn how to use chemical fertilizers intelligently and accurately.

Chemical fertilizers used to feed crops can be compared to the foods we eat. Some promote health. Others, alone or in various combinations, undermine and destroy health.

Fertilizers act similarily in plants. Feed them intelligently and properly and the plants will be healthy and productive. Feed them carelessly, or give the wrong combinations, and they perform poorly and may even die!

All extremes, both good and bad, affect the productivity and the normal functioning of the life processes and result in sickness and unproductivity.

Man can regulate 13 of the 16 essential chemical fertilizers which plants require, and a proper balance of these fertilizers is as necessary to the productive performance of plants as is a proper balance of fruits, grains, nuts, and vegetables to man's health and productivity.

And in addition to the essential nutrients, the plants require water, exercise, anchorage, sunlight, warmth, living space, and protection from extremes in weather, from disease, and from insects. This partial list emphasizes that the success or failure of crops is largely influenced by the soils they grow in.

Unfortunately, most of the tillable land around the world today sets so hard that plant roots need to be equipped with claws, or jackhammer jaws, to penetrate them. Since this is the case, more information on soil management should be shared.

This section points out that tomato plants may fail to produce a crop because of the soil they are grown in, or because some essential fertilizer nutrient is missing, or that the nutrients are not in proper balance.

A deficiency in even *one* essential nutrient can result in crop failure. For example, for a lack of just 20 pounds of boron per acre, if the nutrient is deficient, a crop can fail to mature.

A bag of complete fertilizer.

22

Insects And Soil Maggots

Insects

Many kinds of worms and other insects love tomatoes. Butterflies lay eggs on tomato vines in the daytime and millers (night-flying butterflies) lay eggs on the plants at night.

Some kinds of worms and insects damage the leaves only. Other worms and flies concentrate on spoiling the fruit. And still others affect only the roots.

Many people are surprised to learn that there are more insects active during the night hours than are seen in the daytime.

Insects multiply rapidly and, if not detected early, they can inflict considerable loss to crops.

A recommended program for keeping insects under control and to hold crop losses to a minimum is to implement a regular spray or dusting program (either one is effective) every 7 to 10 days.

Safer and improved sprays and/or dusts, to combat worms and other insects, are being introduced to the public rather frequently. Therefore, it is not prudent to make specific product recommendations. The local agricultural officer is your friend and he has information on the best and safest materials to use.

Hornworm.

Tomato fruit worm.

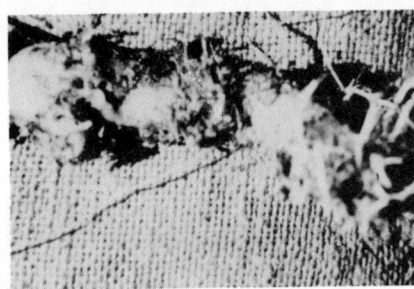

Soil maggots.

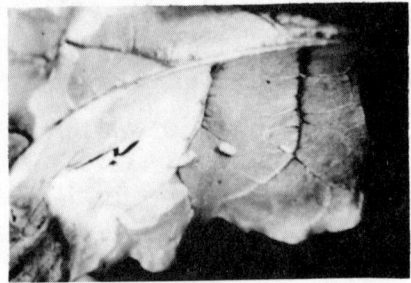

Leaf miners.

Other Factors To Look At When The Fruit Drops

When tomato plants set fruit and then within a few days the tiny tomatoes drop off, those tiny fruits should be examined carefully. A close examination frequently reveals that tiny green worms have eaten into the stem end or blossom end of the tiny tomatoes. This has killed the tiny fruit and it drops off.

Another insect which can be destructive is thrips. They are very tiny, slender, narrow insects. They enter greenhouses through the cooling system if the air is not filtered properly.

Thrips fly, and they inhabit nearly all farm and garden crops. Their special dessert is pollen!

INSECTS AND SOIL MAGGOTS

If they are allowed to multiply, they will eat the pollen as it ripens and thus interfere with normal pollination. When there is no pollen, the female flowers die and fall off. Getting tomatoes to set fruit may be as simple as controlling the thrips.

Thrips control revolves around a regular 7 to 10-day program of spraying or dusting of the foliage and flowers of the crops.

Thrips are more difficult to control than some insects since they are not eaters of plant parts in the strictest sense—like worms—and they are hard to reach with an insecticide which kills by contact.

The easiest way to control thrips is by using a systemic insecticide. (systemic means the poison in the product is absorbed by the plant juices and is circulated throughout the plant. The pollen, too, carries the poison, and by eating the pollen thrips are poisoned.)

Thrip.

Soil maggots.

Checking For Soil Maggots

Soil maggots are destructive in garden and field. They attack and ruin many kinds of vegetable and ornamental crops. They are especially fond of the brassicas (the cabbage family) and onions.

Fortunately, it is not often that soil maggots attack tomato plants, but don't depend on it! When their favorite crops are missing, they will attack less desirable crops.

Diagnosing For Soil Maggots

Some vegetable crops are vulnerable to soil maggot attack anytime, even into harvest time. Watch for telltale signs of their activity.

Here's how: If all the plants have a deep-green color, and are growing evenly, this is good evidence there are no soil maggots.

If, however, even one, two, or three plants have stopped growing, are dull green in appearance, and slightly wilted, the chances are high that maggots are working. Inspect such plants carefully. This is the way to check:

Remove the soil around the stem.

Inspect for maggots.

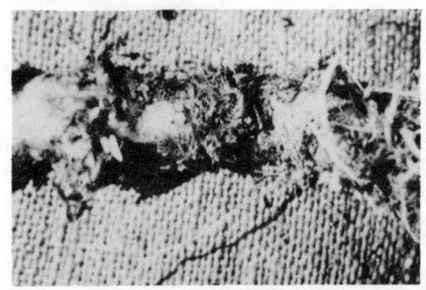

Maggots or disease?

Soil maggots.

Step one: Scratch away the soil that is against the stem of the plant.

Step two: Expose 2 to 3 inches of the stem below the soil surface.

Step three: Examine the exposed stem carefully for areas of decay or for tiny white worms and worm trails, and also for missing bark (the cambium layer).

Soil maggots are tiny white worms, less than one-eighth-inch long and about the thickness of the lead in a lead pencil. The true cabbage maggot has a sharp-pointed black head.

Step four: If the exposed stem looks normal, squeeze it lightly with the thumb and fingers. The object is to try and find a soft spot on the stem.

Step five: If a soft spot is found, remove the plant from the soil—roots, stem, leaves and all.

Step six: Break or cut the soft spot open. If soil maggots are present they will be seen easily. If maggots are not found, the soft spot might indicate that the plant is dying from disease.

Step seven: In either case, to minimize the possibility of spreading either the disease or the maggots, destroy the plant promptly and wash the hands with soap and water.

Healthy roots are clean and white. If soil maggots are found, every plant in the grow-box should be treated with a soil drench to kill the maggots. One soil maggot found on just one plant is proof enough that the entire grow-box or garden area is infested with maggots.

Even though the statement was made earlier that it is not prudent generally to recommend specific products by name to use in controlling insects and disease, there are exceptions. And in this case a specific product and treatment is named and the treatment explained.

Treatment To Control Soil Maggots

If only a few plants or a small area must be treated, dissolve 7 *ounces* Diazinon 45% WP (Wettable Powder) in 30 *gallons water.*

For a larger area and more plants, dissolve 12 to 14 *ounces* Diazinon in 55 *gallons water.*

Mix the contents thoroughly.

Fill a container with the drench and dip ½ to 1 *pint* (depending on the size of the plant) from the container and pour it around the stem at the base of each plant.

A vital spray program.

Soil maggots attack the base of the plant just below the soil surface, and are usually no deeper than 1 inch below the soil surface.

Pour the drench around each plant stem and drench every plant. The Diazinon drench acts quickly and thoroughly. One 55-gallon mixture of drench will treat 420 to 440 plants, at one pint per plant.

The adult maggots are flies similar to the house fly. The females lay the eggs which hatch into soil maggots. Several generations of maggots usually attack in the same season. Thus, it may be necessary to repeat the drench later in the growing season.

23

Plant Diseases

Nearly everyone recognizes that disease can attack plants, and even mild infection can result in serious crop losses.

It is nearly impossible to grow tomatoes without experiencing fungus disease problems. The damper the weather and the higher the humidity, the greater the incidence of fungus attack.

A simple but effective method to reduce fungus buildup is crop rotation. It is a good management practice to rotate crops and plant tomatoes only once in two years on the same land.

Fungus disease grows from spores and spreads rapidly. Its spread is controlled more easily when a spray program is begun before the fungus spreads. Therefore, it is important that spraying with the proper materials is started as soon as the fungus appears.

Frequently, insects and fungus troubles attack at the same time. To control both, only one spray application on a 7 to 10-day interval is necessary. The materials to control both problems are compatable and can be mixed and sprayed together.

Greenhouse Crops An Exception

Rotating crops in the field is recommended. But greenhouse crops are frequently not rotated, and successive crops of the same kind are grown year after year.

Soil sterilization makes this possible, and sometimes the soil is sterilized between each crop.

The determining factor in deciding when to sterilize is the presence of disease. Every time a disease develops the soil is sterilized, using either steam or methyl bromide gas.

"Curley-top" disease.

"Curley-Top" Disease

This disease is quite easy to recognize. The first symptoms appear on the growing tip. The living-green color of healthy plants changes to yellowish-green. The growing tip curls; the youngest leaflets are deformed and the leaves just below the growing tip tend to curl. The older leaves develop yellow and brown areas. Growth is retarded and finally stopped completely. The plant deteriorates and dies.

There is no satisfactory treatment to cure "Curley-top" disease. Thrips and other insects spread the disease from one plant to another.

To minimize the severity of the disease, the infected plants should be pulled up and destroyed promptly and a regular spray program should be implemented to control the insects.

"Curley-top" disease.

"Early and Late Blight" Disease

These are serious diseases and can invade a growing crop any time, but usually infection occurs after the plants are carrying fruit.

"Early Blight" affects the leaves but not the tomatoes. In this respect it differs from "Anthracnose" disease which affects both the leaves and fruit, and is especially serious on beans.

The symptoms of "Early Blight" disease are brown-to-black depressed spots on the leaves, with gray, thin whiskers around the perimeter of the sunken spots. The gray whiskers are the spores (seeds) by which the disease spreads.

"Early Blight" attacks both the old and the young leaves. The sunken spots on the leaves vary in size from mere dots to one-half inch or larger.

"Late Blight" disease has the characteristic symptoms of "Early Blight" in the early stages of infection. The disease spreads rapidly through a crop of tomatoes. As infection progresses, black lesions develop along the midribs of the leaves and along the stems. Dark brown-to-black spots develop on all parts of the plant, including the fruit. Unless the disease is arrested, the spots on the tomatoes will become watery, and the tomatoes will rot and eventually fall off. The black lesions on the stems and midribs of the leaves increase in size and penetrate the living tissues. Later the stems become watery, decay, and the plant dies.

Early Blight.   *Late Blight.*

Treatment: Both "Early" and "Late Blight" disease can be controlled sufficiently to avoid losing an entire crop, providing the right materials are used promptly, regularly, and accurately.

But the use of fungicides to control disease should be regarded as palliative measures only. The only lasting and satisfactory procedure is to use heat-treated, disease-free seed and plant only healthy plants in sterilized soil.

For complete information on greenhouse sanitation and disease prevention and control practices, refer to chapters 58 and 59 in the book *Food For Everyone*, by Mittleider and Nelson.

"Virus" Disease

There is a difference between a "virus" and a "bacterial" disease in plants.

PLANT DISEASES

A "bacterial" disease usually enters a plant from the outside first, often through an injury such as a scratch or bruise, and the infection is generally localized in the early stages. The infection can result in killing the plant, as it often does.

A "virus" disease enters the blood-stream (the plant sap) through many avenues. Once it is in the blood stream it is next to impossible for the farmer or gardener to eradicate it.

"Curley-top," "tobacco-mosaic," and "bacterial-canker" are diseases to be feared with dread by tomato growers.

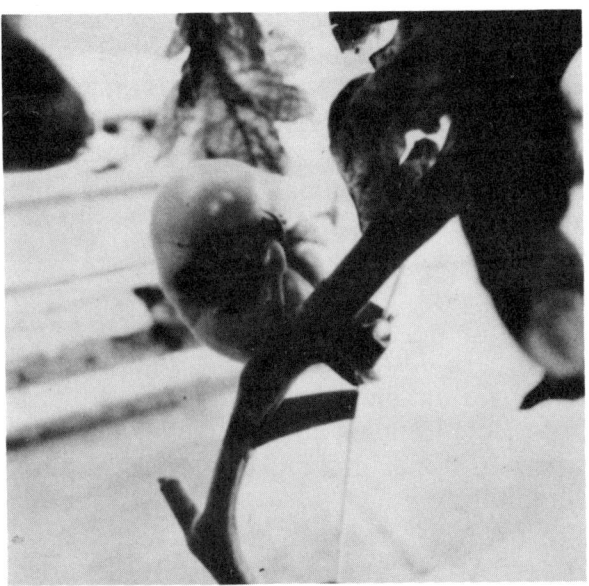

Bacterial canker.

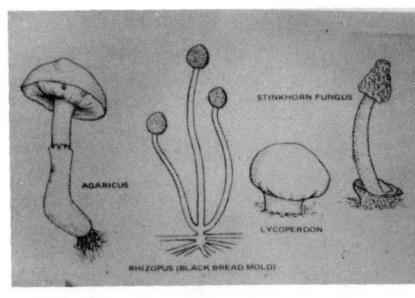

Fungi.

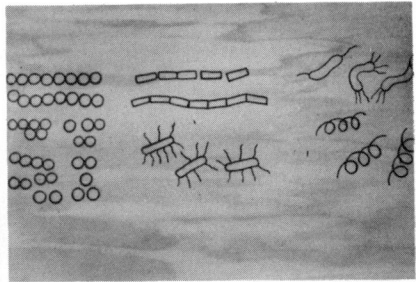

Bacteria.

Virus disease.   *Tobacco Mosaic.*

Treatment: Plants which are infected with "tobacco-mosaic" virus disease lose the dark-green color of healthy plants. The plants practically stop growing. The terminal bud cells and small leaflets are deformed. The fruit has broad brown streaks in the shoulder. The surface of the *stem end* of the tomatoes is rough. The *shoulder* is sunken and corrugated in appearance.

All infected plants should be destroyed promptly, preferably by burning.

Select tomato varieties that are resistant to "mosaic" disease, and maintain clean and sanitary premises to prevent "mosaic" infection. Request all who handle and smoke tobacco not to touch and not to handle tomato leaves, or to smoke around your plants.

Important: After handling plants that are not really healthy, wash the hands with soap and water before touching other plants or handling sterilized soil.

"Bacterial" and "Virus" diseases are spread to healthy plants just by touching their foliage after handling a diseased plant.

24

Weather Problems And Vibrating Plant Vines

When plants fail to set fruit, the amateur grower often points to the weather as the cause. This is the reason for the question, "Does temperature affect the fruit-set on tomatoes?" The answer is, of course, "Yes!"

Prolonged cool and cold weather will slow down the growth of tomato plants and the fruit will grow very slowly in cold weather. However, healthy plants flower, pollinate, and set fruit even in cold weather.

Tomatoes are considered warm-weather plants. They perform best in temperatures that are neither cold nor hot, between 75° and 95°F.

During prolonged temperatures above 100°, tomato plants practically stop growing. And if the vines show wilting during the middle of the day, it is possible for the flowers to fall off after pollination. This would result in poor to zero fruit set.

The correction in such cases is to provide diffused light, through shading, to lower the temperature to 95° or less, and to supply adequate water to keep the plants from wilting.

Shaking and Vibrating Tomato Vines

For some greenhouse growers it is a daily routine practice to vibrate each tomato vine, or shake the overhead wires which hold the vines. Apparently this is to pollinate the flowers. Just when and how this practice began would be interesting to know.

The merits and validity of the operation can be challenged easily from other greenhouse operators whose crops always set fruit automatically. They do not vibrate or use any other palliative practices to pollinate tomato flowers.

Summary

To summarize this detailed discussion on the tomato fruit-set problems, the list of possible factors include

- Soils
- Improper balance or a lack of one or more essential fertilizer nutrients
- Poor light or extremes in temperatures
- Insects
- Disease
- Weather, etc.

The two most likely causes are soils and fertilizers. The grow-box method makes it simple and easy to control both of these.

Experience bears out that when the grow-box method is followed carefully, pollination and fruit set occur automatically.

A superior yield.

25

Cracked Fruit

Every grower seems destined to have some cracked tomatoes, and some varieties produce more cracked tomatoes than others, even when grown under the same conditions.

Growing two or three varieties together is good management practice, and will help determine which varieties do best for specific conditions.

It should be recognized, however, that excess cracking and splitting are symptoms of some disorder or problem. For example, poor watering.

Diagnose the reason for cracked tomatoes.

The available water should be ample and uniform at all times. Plants loaded with tomatoes require large amounts of water continuously. Applying water on warm days stimulates rapid growth. A reduction in the available water supply results in reduced growth.

The aim in watering should be to supply a *uniform* amount of available water at all times. This practice encourages uniform growth day after day and is necessary to produce well-shaped fruit free from cracks. Adequate drainage and watering every day, even twice a day, is recommended for plants loaded with tomatoes.

Another factor which can influence excess cracking of the tomatoes is fertilizers. If cracks occur on the blossom-end of the fruit, or the fruit has sunken shoulders, it frequently indicates a nutrient deficiency.

Cracks which develop on the *stem-end* can also be from a nutrient deficiency, or from improper watering.

The most commonly occurring nutrient deficiencies and the specific amounts of the nutrient compounds required to correct deficiencies are given in Appendix 1. If nutrient deficiencies do occur, follow the instructions carefully and make the proper corrections promptly.

Deficiency corrections can be made for each of the essential nutrients. If a deficiency *does exist,* the corrective treatment is adequate, and if a deficiency *does not* exist, applying the corrective treatment *will not* produce harmful affects on the crop.

26

Blossom-End Rot On Tomatoes

There are two kinds of blossom-end rot. One is caused from stress in the plant. The other is caused from a fungus infection. The two have different characteristics.

Blossom-End Rot Caused From Stress

Blossom-end rot caused from stress is frequently due to the leaves wilting and, as a result, sap is being pulled from the fruit. Each of the following stress factors can cause blossom-end rot:

- Low calcium
- Low potassium
- Low boron
- Low nitrogen
- Low water
- High-soluble salts in the water

Blossom-end rot caused from stress may occur under *any condition* which limits the moisture level in the plant.

Blossom and Stem-End Rot Caused From Infection

During cool, damp weather, or high humidity, fungus spores may infest the stigma of the tomato flowers during the hours they are receptive to pollen.

After pollination, the flower petals collapse over the stigma and this seals off air circulation over the stigma. Without air circulation over the stigma, it stays damp and the fungus spores and tiny tomatoes begin to grow together. The fungus spores penetrate the blossom-end of the tiny tomatoes through the style of the stigma.

As the tomato grows, so does the fungus inside the fruit. Possibly 14 days later, the infected tomatoes are ruined. If the fungus is allowed to spread, the watery fungus area enlarges till the tomato becomes slime.

Although there are two kinds of blossom-end rot on tomatoes, they are different and can be easily identified.

Blossom-end rot caused from stress is usually shallow, brown-blackish in color. It remains localized at the blossom-end, and is *not* watery or slimy. It usually remains shallow (this depends on the severity of the stress factor) and often does not enlarge, or penetrate, deep into the fruit.

Blossom-end rot caused from fungus infection is dark brown and black in color. It is wet and slimy, and keeps spreading, both on the outside and in the inside of the tomato fruit.

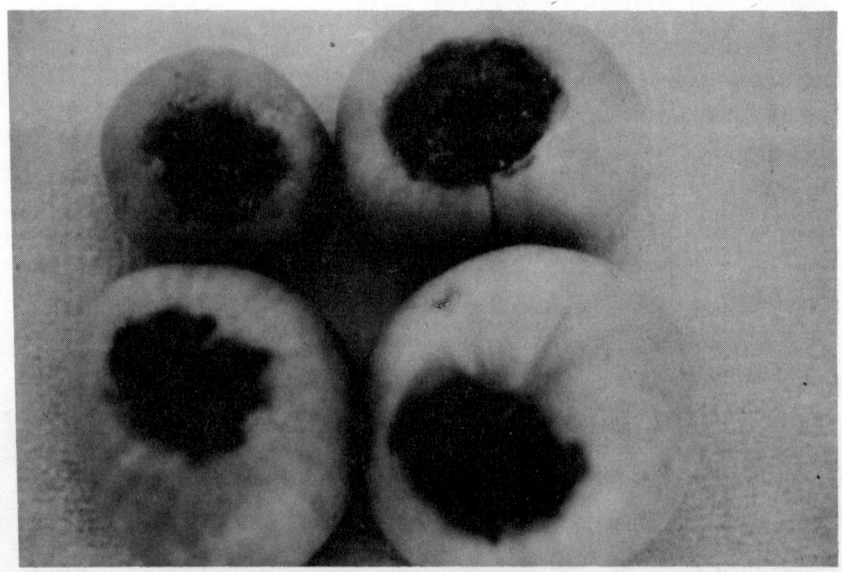

Blossom-end rot.

BLOSSOM-END ROT ON TOMATOES

If you have had experience with "Athlete's Foot," a fungus disease, you know how it spreads and penetrates. Blossom-end rot fungus disease in tomatoes spreads and penetrates in a similar way.

Either case of blossom-end rot disease can be controlled.

If the cause is from a fungus infection, the crop can be saved by implementing a regular spray program of control early. The recommended treatment is to spray the foliage with the proper fungicide when the disease is first recognized and continue with repeated spray applications every 7 to 10 days until the crop is harvested.

If the cause is from wilting due to stress within the plant, the crop can be saved by eliminating the cause of the stress factor.

Moisture stress.

27

Thinning Tomatoes

It may seem ironic to follow an extended discussion on fruit-set problems with a section on thinning the flower clusters, but keep in mind that thinning is becoming more popular year after year.

Premium prices are paid for tomatoes that are well-shaped, uniform, and of a specific size.

Experience bears out that thinning to 4 or 5 tomatoes per flower cluster (hand) *does not* noticeably reduce the total pounds of fruit the vine will produce.

Thinning tomatoes is similar to thinning fruit trees. It increases the size of the fruit and also improves the shape and quality of the remaining crop.

Some tomato varieties should not be thinned. These include cherry tomatoes, pear and "Pixey" tomatoes, and field tomatoes grown for canneries.

The plants grown for thinning are first pruned to single stem vines and are tied to stakes as they grow, or are guided around strings which are tied to overhead wires.

Pruned single-stem plants develop flower hands which have from 4 to 50 flowers per hand, depending on the variety. The hands occur 6 to 10 inches apart on the stem.

It seems that each hand of tomatoes has the potential to produce a specific weight of ripened fruit, whether there are 5 tomatoes or 10 fruit per hand. The difference is recognized in the size of the tomatoes. In other words, if one hand is thinned to 4 or 5 tomatoes, and another hand is allowed to mature 8 or 10 fruit, the total weight of tomatoes in each hand will be nearly the same.

Obviously, the hand with the least number of fruit will produce the larger-size tomatoes. This is the reason thinning to 4 and 5 tomatoes per hand is practiced.

28

Harvesting Tomatoes

Every grower lives in fond expectation, looking forward to harvest time. For this he dreams and works throughout the growing season. If he has done his part well he has good reason to expect a bountiful harvest.

Whether the crop is sold to markets, or whether it is grown for home use, are factors which help decide when to harvest tomatoes.

Markets want tomatoes with 2 to 4 weeks shelf life. Therefore, market-bound tomatoes are picked when the green color first changes to cream-color.

Large yield.

Leaving tomatoes on the vines until red-ripe develops more cracked tomatoes, but the bonus in flavor of vine-ripened tomatoes far surpasses the inconvenience caused from the extra splitting.

Shipping and market trends discourage the production of vine-ripe tomatoes for the supermarket trade. Therefore, for much of the world's younger population, it is probably true that they seldom or never taste the goodness of a vine-ripened tomato, fresh off the vine! Even in America, the majority who enjoy the real goodness of tomatoes are those who grow them.

Red-ripe tomatoes crack easily when picked. Therefore, ripe tomatoes should be handled gently to avoid bruising and unnecessary splitting.

It is clear that harvesting of tomatoes can and does vary. And fortunate is the family that can enjoy fresh vine-ripe tomatoes, even if this is possible for only a few weeks of the year.

Canned vine-ripened tomatoes are a good second choice throughout the rest of the year.

Lengthening the Ripe-Tomato Harvest Season

Don't let an early frost cut your picking season short! Here's how to lengthen the tomato season:

1. By following the information outlined in this book and growing your own plants from seed, you can add weeks to the harvest season. But even so, tomatoes are everbearing and the vines will very likely still be loaded with fruit when winter strikes.

2. Don't let the crop freeze! Just before the frost strikes, pick the green tomatoes and wrap each one separately in newspaper, and store them in a cool place. Or, if possible, place the green tomatoes in single layers (unwrapped) on shelves in a cool pantry or store room.

The green tomatoes will ripen very slowly. It is possible to store them for two or three months and keep eating red tomatoes all the time.

In the Northwest, and other colder sections of America, the first killing frost strikes about the middle of September and lasts 2 or 3 nights only. After this first freeze, the weather is warm again for several weeks before the general freezing cold sets in for good.

It is worth the effort to cover the tomato vines where they are growing to protect them from the first early frost. This will extend the picking season several weeks.

Later, when it is obvious that the freezing weather has come to stay, where possible, do the following:

Step one: Stretch a strand of #8 wire 7 or 8 feet high inside the car garage.

Step two: On the evening before the first severe frost, pull the tomato vines, roots and all.

Step three: Shake the soil from the roots, but *do not* pick the tomatoes off the vines.

Step four: Carefully cut the strings which hold the plants.

Step five: Take the vines, the leaves, the roots, and the tomatoes into the garage and drape each plant over the wire.

Step six: The temperature in the garage should not fall below 32°F., or the tomatoes will freeze.

Green tomatoes hanging on the vines and stored in an unheated garage will continue to ripen for 8 to 12 weeks. The fascinating results are that tomatoes hanging from the vines (over the wires) seldom rot, and red tomatoes can be picked from the vines into early January without any additional expense.

Considerable effort has gone into preparing this publication. The driving force has been to help more people get better acquainted with the world's most popular vegetable, and be able to enjoy its health-giving properties over a longer season yearly. Now it's yours to enjoy more assorted vegetables and vine-ripened tomatoes!

More fruit coming!

Appendix 1
Nutrient Deficiencies, Symptoms And Corrections

Nitrogen Deficiency

Symptoms: General yellowing over entire plant; spindly, stunted growth.
Correction: Two pounds ammonium nitrate (34-0-0) per grow-box.

Phosphorus Deficiency

Symptoms: A purplish discoloration on older leaves; stunted growth; poor fruit set.
Correction: One pound di-ammonium phosphate (18-46-0) per grow-box.

Potassium Deficiency

Symptoms: Scorching (firing) of edges of mature leaves; shriveled seeds in cereal crops; poor fruit quality.
Correction: One-and-a-half pounds potassium sulfate or chloride per grow-box.

Magnesium Deficiency

Symptoms: Dead areas in older leaves; tend to produce bright colors; older leaves die from edges inward.
Correction: Two pounds 12 ounces magnesium sulfate (epsom salt) per grow-box.

Calcium Deficiency

Symptoms: Dead terminal buds; stunted growth; poor root growth.
Correction: One pound 12 ounces calcium nitrate per grow-box.

Iron Deficiency

Symptoms: Yellowing of interveinal leaf tissue while veins remain green.
Correction: One pound 4 ounces iron sulfate per grow-box.

Boron Deficiency

Symptoms: Black heart of tubers; death of terminal buds.
Correction: Two ounces (60 grams) boron (sodium borate) mixed with 3 quarts sawdust or sand.

Molybdenum Deficiency

Symptoms: "Whiptail disease," narrow long leaves, producing twisted pattern.
Correction: Fifteen grams (one-half ounce) sodium molybdate or molybdic acid, mixed in one cup sawdust or sand per grow-box.

Appendix 2
Fertilizer Formulas

Preplant Fertilizer Formula

    6   pounds phosphate (18-46-0)
    4   pounds potassium, either sulfate or muriate of potash
    4½ pounds ammonium nitrate or 7 pounds sulfate of ammonium
    4½ pounds magnesium sulfate (epsom salt)
    4   ounces boron sodium dorate or boric acid

Spread separately:

    <u>11</u>   pounds lime (see note on rainfall, page 41)
    30   pounds total

Mittleider Weekly Feeding Formula

    9   pounds calcium nitrate
    4   pounds ammonium nitrate
    1½ pounds di-ammonium phosphate (18-46-0)
    4½ pounds potassium sulfate or chloride
    6   pounds magnesium sulfate (epsom salt)
    8   *ounces* iron sulfate
    4   *grams* copper sulfate
    8   *grams* zinc sulfate
    12 *grams* manganese sulfate
    12 *grams* boron (sodium borate or boric acid)
    <u>3</u>   *grams* molybdenum (sodium molybdate or molybdic acid)
    25½ pounds total

Constant Feed Solution

    55 gallons water
    1   pound Weekly Feeding Formula (above)

Note: This solution can be used for every watering

Appendix 3
Units Of Measure

48 teaspoons .... 1 cup
60 drops .... 1 teaspoon
3 teaspoons .... 1 tablespoon
1 tablespoon .... ½ ounce
16 tablespoons .... 1 cup
1 cup .... 8 ounces
16 fluid ounces .... 2 cups
2 cups .... 1 pint
½ liquid pint .... 1 cup
2 pints .... 1 quart
4 quarts .... 1 gallon
1 pound .... 16 ounces
1 pint .... 1 pound
1 gallon .... 8.337 pounds (8 pounds)
1 mile .... 5,280 feet or 320 rods
1 acre .... 43,560 square feet or 160 square rods

To Change Centigrade To Fahrenheit

Multiply centigrade by 9/5 and add 32 degrees.

To Change Fahrenheit To Centigrade

Subtract 32 degrees and multiply by 5/9.

Equivalent Rates In Applying Fertilizers

1 ounce per square foot .... 2,722.5 pounds per acre
1 ounce per square yard .... 302.5 pounds per acre
1 ounce per 100 square feet .... 27.2 pounds per acre
1 pound per 1,000 square feet .... 43.6 pounds per acre
1 pound per acre .... ⅓ ounce per 1,000 square feet
5 gallons per acre .... 1 pint per 1,000 square feet
100 gallons per acre .... 2½ gallons per 1,000 square feet
100 gallons per acre .... 20 pounds per 1,000 square feet
100 gallons per acre .... 1 quart per 100 square feet

Index

A
"A"-Frame Method—64-70
Anthracnose disease—121

B
Bacterial disease—122-124
Blossom-end rot—causes of, 129-131

C
Constant Feed solution—watering seedflats with, 43; watering transplanted tomato shoots, 47; formula for, 48; how to apply, 49-50; used every watering, 58
Cracked fruit—causes for, 127-128
Curley-Top disease—120-121

D
Dibble—how to make, 45
Diazinon—117-118
Disease—treating seeds against, 36-37; light important in controlling, 72; keeping stems dry prevents, 81; explanation of various, 119-124
Dusting program—113

E
Early Blight disease—121-122

F
Fertilizer—preplant formula, 39-40; weekly feeding procedure, 93-98; on transplanting day, 98-99; nitrogen, 99; when to stop applying, 100-101; problems with, 110-111, 128; formulas, 139; equivalent rates in applying, 140
Field capacity—term for soil water saturation point, 90
Flowers—8 weeks from seed planting, 85
Flower dusters—82
Flower set—first flower buds, 62; solving problems with, 102-106

Food for everyone—references to, 37, 76, 122
Fungus infection—130-131

G
Germination—reduced by hot water treatment, 36; fertilizer applied to unsprouted seeds delays, 42
Grow-Boxes—described, 20; constructing, 21-29; soils for, 30-32
Gypsum—use in low rainfall areas, 41

H
Hanging vines—at end of season, 134-135
Harvesting—133-135
Heat—ideal for tomato production, 103
Hornworm—114
Hot water—treating seeds with, 36-37
Hygroscopic—tendency of preplant formula components, 40

I
Infection—129-131
Insects—113-118

K
Kreosote—toxic to plants, 24

L
Late Blight disease—121-122
Leaf miners—114
Leaves—of vigorous plant, 63; oldest leaves turn lighter color, 81; diagnosing leaf symptoms, 95, 116
Lengthening harvest—134-135
Light—controls position for grow-boxes, 73, 103
Lime—omit in pre-packaged formula application, 40; types to use in low and high rainfall areas, 41

Location—for gardens, 21-22
Love-apple—name for tomato, 18
Lycopersicon—family name for tomato family, 18

M

Mixing crops—in grow-boxes, 72
More Food From Your Garden—references to, 29, 74

N

Nematodes—107-109
Nutrient Deficiencies—137-138

P

Plant population—104
Planting—in grow-boxes, 65; positioning crops in relation to sunlight, 73; in fields, 104; in greenhouses, 105
Pruning—temporarily stops upward growth, 52; don't remove terminal bud, 57; schedule for pinching, 57; removing suckers, 60; pruning fruiting tomatoes, 79-86

R

Rotation—annual, 119

S

Seeds—choosing correct s. for your area, 33-35; advantages to growing from s., 34-35; facts to know about s., 36-39; starting plants from s., 39-43; number per flat, 41; time for transplanting 56
Soil air—needed along with water, 90
Soil maggots—113-118
Soils—grow-box, 30-32; materials for custom-made, 32; sterilizing, 38; problems in regular, 38, 110-111

Solanum—tomato part of s. family, 18
Spacing—in grow-boxes, 65, 71-72
Spindly—preventing plants from becoming, 51-53
Spray program—113
Staking—instructions for, 61-62, 66
Sterilizing soil—in oven, 38; disease requires, 120
Suckers—pruning of, 60, 83

T

Temperature—best tomato-growing, 51-52, 103, 125
Thinning—132
Thrip—114-115
Tobacco-Mosaic diseases—123-124
Tomato fruit worm—114
Transplanting seedlings—into pots, 44-47; into larger containers, 54-59; into grow-boxes, 58, 72, 75-78
Tying—instructions for, 61-62, 68-69; weekly process, 87-88

U

Units of measure—140

V

Vibrating plant vines—125-126
Virus disease—122-124

W

Watering—never use fertilizer solutions on unsprouted seeds, 42; water flats daily, 52; when transplanting, 54-55; daily, 89-92; aim in watering, 128
Weeds—control difficulties in regular gardens, 38-39
Weekly Feeding—78; formula, 48
Wilt—tomatoes slow to, 92